SpringerBriefs in Electrical and Computer Engineering

Nauman Khan • Soha Hassoun

Designing TSVs for 3D Integrated Circuits

 Springer

Nauman Khan
Department of Computer Science
Tufts University
161 College Ave
Medford
USA

Soha Hassoun
Department of Computer Science
Tufts University
161 College Ave
Medford
USA

ISSN 2191-8112   ISSN 2191-8120 (electronic)
ISBN 978-1-4614-5507-3   ISBN 978-1-4614-5508-0 (eBook)
DOI 10.1007/978-1-4614-5508-0
Springer New York Heidelberg Dordrecht London

Library of Congress Control Number: 2012946733

Printed on acid-free paper

Springer is part of Springer Science+Business Media (www.springer.com)

To my wife and my daughter.

NHK

To my parents, who gave me life, and my children, who give it light.

SH

Preface

Stacking multiple dies to form 3-D integrated circuits (ICs) has emerged as a promising technology to reduce interconnect delay and power, to increase device density, and to achieve heterogeneous integration. Through-silicon vias (TSVs), metallic interconnect between dies, are a key enabling technology for 3-D ICs. TSVs can be used for routing signals, for power delivery, and for heat extraction. TSV manufacturing advances are well under way. However, there is little experience in designing optimally with TSVs.

This book explores challenges and best strategies to design with TSVs and offers several key contributions. Signal TSVs induce noise in the substrate and affect neighboring devices. This book proposes a novel technique, the GND Plug, to mitigate TSV-induced noise. The results show the superiority of this technique in grounding noise compared to other techniques adapted from 2-D planar technologies such as a backside ground plane and traditional substrate contacts. The book also investigates, in the form of a comparative study, the impact of TSV size and granularity, spacing of C4 connectors, off-chip power delivery network, shared and dedicated TSVs, and coaxial TSVs on the quality of power delivery in 3-D ICs. The book provides detailed best design practices for designing 3-D power delivery networks. TSVs occupy silicon real estate and impact device density. This book provides four iterative algorithms to minimize the number of TSVs in a power delivery network. Unlike prior work, these algorithms can be applied in early design stages when only functional block-level behaviors and a floorplan are available. Finally, the book explores using carbon nanotubes to design the TSVs and power grid, and the results show that the use of carbon nanotubes for grid design offers substantial advantages in terms of reducing IR drops. Overall, the book advances 3-D IC design.

We are grateful to Syed Alam, senior member of technical staff at Everspin Technologies in Austin, TX, who provided inspiration, technical wisdom, and support to Nauman during his PhD thesis. We are also grateful to Sherief Reda, associate professor at Brown University, for his contributions to the early estimation algorithms.

Contents

Chapter 1
Introduction

Moore's law has inspired the growth of integrated circuit (IC) technology since its inception in 1965 [75]. Each new technology node produces smaller and faster devices keeping pace with Moore's prediction of $2\times$ scaling every 18 months. The exponential decrease in feature size, from $10\,\mu m$ [87] to $22\,nm$ [20] over the past four decades, has resulted in an astronomical performance increase. For this trend to continue, significant challenges need to be overcome in several key areas [74]. IC technology has evolved from a device-centric technology to one where interconnect also plays a critical role. The latency of interconnect dominates that of transistors [70]. Oxide thickness of a metal oxide semiconductor field effect transistor (MOSFET) determines the size and the leakage current of a transistor. Oxide thickness approaching atomic levels imposes a practical bound on the leakage current and hence limits transistor sizes [93, 105]. Exponential increase in capital cost, to set up a foundry, poses a threat to the viability of future technology scaling [26].

Vertical stacking of dies, forming a 3-D IC, is a promising alternative to traditional 2-D ICs to keep pace with Moore's Law. 3-D IC technology is not only capable of increased device-density but also offers heterogeneous integration of dies from disparate technologies (analog, digital, mixed signals, sensors, antenna, and power storage) and from different technology nodes (65, 32, 22 nm etc.). The 2009 International Technology Roadmap for Semiconductor (ITRS) predicts that by 2015 the number of stacked dies in memory, low-end low-cost, and high-performance chips will be 9, 3, and 2, respectively.

1.1 More than Moore with 3-D IC Technology

The concept of 3-D integration is not new. Although first introduced in 1960 [37], 3-D integration did not materialize until recently because of two reasons: (1) 2-D ICs were already producing results keeping pace with Moore's law, and (2) technology

N. Khan and S. Hassoun, *Designing TSVs for 3D Integrated Circuits*, SpringerBriefs in Electrical and Computer Engineering, DOI 10.1007/978-1-4614-5508-0_1, © The Authors 2013

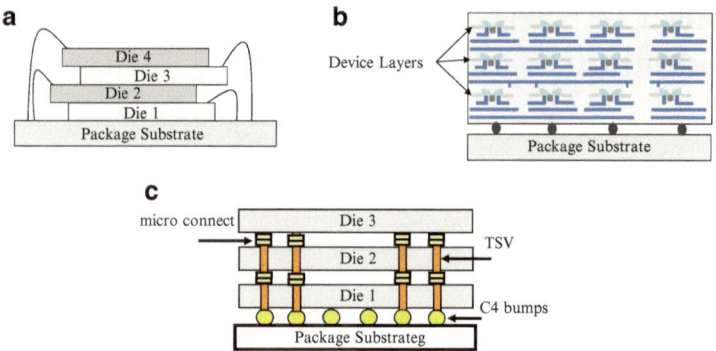

Fig. 1.1 3-D integration techniques. (**a**) Wire-bonding, (**b**) Monolithic 3-D IC, (**c**) TSV integration

to remove the tremendous amount of heat from an IC was not available. A combination of technical innovations to produce ultra-low power devices [92] and of future scaling challenges has turned the focus back to 3-D ICs [74].

3-D stacking provides exciting design opportunities not possible with 2-D ICs. Various levels of granularity exist to partition a system to form a 3-D system. Fabricating individual dies using various technologies and stacking them to form a 3-D IC is an example of coarse granularity [84]. Monolithic 3-D integration, stacking two dies such that one die contains all the NMOS transistors and the other contains all the PMOS transistors, is an example of the finest granularity [40, 73]. Loh et al. provide a detailed architectural analysis of these granularities [66]. The benefits are many. At the circuit-level, the length of global wires can be reduced by as much as 50%, wire-limited clock frequency can be increased by $3.9\times$, and wire-limited area can be decreased by 84% [50]. Power requirements can be reduced by 51% at the 45-nm technology node [11]. At the architecture-level, optimal 3-D design can be achieved by co-optimizing the architecture and technology at each design stage [39, 65].

Techniques to achieve 3-D stacking include wire-bonding, monolithic integration, and through-silicon vias (TSVs). While straight-forward and economical, wire-bonding is limited to low-power low-frequency ICs that require less inter-die connections. Wire-bonding may occur for staggered dies, shown in Fig. 1.1(a). A monolithic 3-D IC, shown in Fig. 1.1(b), requires a new semiconductor manu-facturing technique where multiple layers of devices can be fabricated in a single substrate. This technique offers the most stacking at the highest cost. TSV-based 3-D integration, shown in Fig. 1.1(c) and the focus of this book, is a promising technique to achieve short and dense inter-die interconnects (an order of magnitude higher than wire-bonding). Inter-die communication can also be achieved by inductive coupling [47]. While complex, expensive, and requiring new design and fabrication solutions, TSV-based design has gained the most momentum.

The IC industry has enthusiastically embraced 3-D stacking technology, and progress is being made to advance design methodologies, tools, and core technologies. Qualcomm Inc., the biggest near-term consumer of 3-D ICs for mobile devices, is working on standardizing the process flow [39]. All major EDA vendors are developing design tools to support each stage of the 3-D IC design cycle. MIT Lincoln Labs has introduced a fully depleted silicon-on-insulator (FDSOI) process that can stack three dies using wafer–wafer oxide bonding [2]. Tezzaron Inc. has demonstrated a variety of 3-D ICs (memory, FPGA, and mixed signal) using a 180 nm technology node and wafer-level, via-first, and metal-to-metal thermal bonding [84]. Foundries are working in close relation with research organizations to develop fabrication processes that meet 3-D stacking requirements [27, 51, 86, 96].

1.2   TSV-Based 3-D IC Design Challenges

TSVs offer higher interconnect density and better performance than wire-bond stacking technology [10]. However, fabricating vertical interconnects that pass through dies containing substrate, devices, and interconnect, poses manufacturing and performance challenges. A few of these challenges are discussed in this section.

TSVs occupy valuable silicon real estate. As a point of reference, ITRS predicts the minimum TSV area to realize global-level inter-die connections to be $16 \mu m^2$ [5], whereas the area of a 6T SRAM cell for 45 nm Hi-K Metal-Gate technology is $0.346 \mu m^2$ [19]. TSVs thus create blockages ($\approx 46\times$ the SRAM cell area): neither a device can be fabricated nor a signal can be routed through the area occupied by a TSV. TSVs therefore impact device density, die floor-plan, and interconnect routing.

Manufacturing constraints associated with TSV etch and via filling processes dictate TSV size [38]. Smaller TSV sizes are desirable but require the silicon (Si) substrate to be thinned to a thickness of 100 to $10 \mu m$, or even less in bulk CMOS technology [76]. To fabricate a TSV of size $5 \mu m$ assuming a practical aspect ratio of 10:1, the maximum die thickness will be $50 \mu m$. Technical innovations are needed to manufacture and bond thin wafers [86].

A TSV is a metallic, usually copper (Cu), interconnect extending through the substrate and insulated by a dielectric material. Any signal transitions through a TSV create noise within abutting substrate. This TSV-induced substrate noise poses a major threat to the performance of neighboring devices and neighboring TSVs. TSV-induced substrate noise increases leakage current, which increases static power consumption and can turn transistors in the "off" state to the "on" state and vice versa [91].

The large mismatch between the coefficients of thermal expansion (CTE) of metallic TSV ($17.5E-6/°C$ for Cu) and Si substrate ($2.5E-6/°C$) results in serious reliability concerns [94, 115]. High temperature loadings during fabrication create thermal stress in the substrate, which in turn impacts the mobility of carriers and affects device performance [95].

Power delivery design for 3-D ICs is a challenging task due to increased device density and package asymmetry [46, 117]. While the die closest to the package can get power supply directly from the package, TSVs are required to deliver power to the dies further away from the package. TSV size and placement are design decisions. Technology and design choices must be explored to minimize TSV area penalty and meet power requirements.

Thermal management is one of the biggest problems faced by 3-D IC design. TSVs can be used to extract heat from the dies away from the heat sink [117]. Optimizing the number and placement of TSVs to meet the thermal budget requires accurate modeling and novel algorithmic and design approaches [120].

1.3   Contributions

The work proposed in this book is original and innovative as it is one of the first to explore design issues concerning TSVs in 3-D ICs. The key contributions are:

- Characterizing the impact of TSVs on substrate noise using a simple circuit model and a three-dimensional finite element solver.
- Proposing an effective technique, the GND plug, to mitigate TSV-induced substrate noise, and comparing the performance and area penalty of the GND plug with other noise mitigation techniques including thicker dielectric liner and backside ground plane.
- Performing the first comparative study of system-level power delivery for 2-D and 3-D ICs utilizing realistic workloads, and investigating methods for achieving 2-D-like power quality in 3-D ICs.
- Studying the use of coaxial TSVs for power delivery and assessing the benefits in terms of the number of routing blockages, decoupling capacitance, and sharing of signal and power routing.
- Investigating algorithmic solutions to estimate early in the design cycle the number of TSVs required for power delivery network design.
- Exploring using carbon nanotubes (CNTs), as an alternative to Cu, for on-chip power delivery and assessing their performance.

The contributions proposed herein pave the way for 3-D IC design using TSVs and motivate further research into this emerging technology.

1.4   Organization of Book

This book consists of several chapters. Chapter 2 provides an overview of TSV technology presenting various technology and modeling aspects of TSVs. Chapter 3 characterizes the impact of TSVs on substrate noise utilizing a three-dimensional finite element solver to extract parasitics. A practical and effective noise mitigation technique, the GND plug, is proposed and compared with traditional noise mitigation techniques.

Designing a power delivery network for a 3-D IC is a challenge and is explored in Chap. 4. A comparative study of system-level power delivery for 3-D ICs utilizing realistic workloads is presented. Several 3-D power delivery configurations are evaluated with the goal of understanding major factors that impact the quality of 3-D power delivery network. Coaxial TSVs are also explored for power delivery in 3-D ICs.

Chapter 5 presents an algorithm to estimate TSVs required for on-chip power delivery consistent with supply noise limits. A number of algorithms are proposed to find the minimum number of TSVs that deliver power with acceptable IR drops.

Chapter 6 utilizes the architectural study presented in Chap. 4 and the area estimation technique presented in Chap. 5 to explore using Carbon nanotubes (CNTs) as an alternative to Cu. CNTs have emerged as a competitive technology for interconnect. A practical CNT model is utilized to quantify electrical benefits of using CNT-based on-chip power grid and TSVs over Cu-based TSVs.

Chapter 7 summarizes the book and outlines future research challenges in designing TSV-based 3-D ICs.

Chapter 2
Background

Through-silicon vias (TSVs) connect multiple dies within a 3-D IC. TSVs can be used to route inter-die signals, deliver power to each die, and extract heat from dies further away from the heat sink [59, 63, 102]. A TSV, shown in Fig. 2.1, is a metal interconnect that passes through the Si-substrate and is electrically isolated from the substrate by a liner formed using an insulating material such as Silicon Dioxide (SiO_2). The liner determines the capacitance of a TSV and is designed to exhibit low leakage current and large breakdown voltage. A barrier layer prevents diffusion of metal from the TSV into the Si substrate. Tantalum (Ta) and Titanium Nitride (TiN) are the most commonly used barrier materials. Cu-based TSVs are preferred due to the lower resistivity of Cu, but Tungsten (W) and Polysilicon TSVs have also been proposed [5].

The rest of the chapter is organized as follows. Section 2.1 provides an overview of different types of TSVs. Section 2.2 describes technical aspects related to TSV integration. Section 2.3 presents a summary of various TSV fabrication techniques. Section 2.4 presents various approaches used to create electrical models of TSVs.

2.1 TSV Types

Over the past few years, a variety of TSV fabrication processes have been proposed resulting in different types of TSVs. Each type has benefits and technological challenges.

2.1.1 Regular (Square or Cylindrical) TSV

The most common TSV shape is cylindrical, as shown in Fig. 2.2, or square. The TSV is made from metals such as Cu, plated Cu, tungsten, or doped Polysilicon (polySi), resulting in a wide range of resistivities. The regular TSV is simpler to

N. Khan and S. Hassoun, *Designing TSVs for 3D Integrated Circuits*, SpringerBriefs in Electrical and Computer Engineering, DOI 10.1007/978-1-4614-5508-0_2, © The Authors 2013

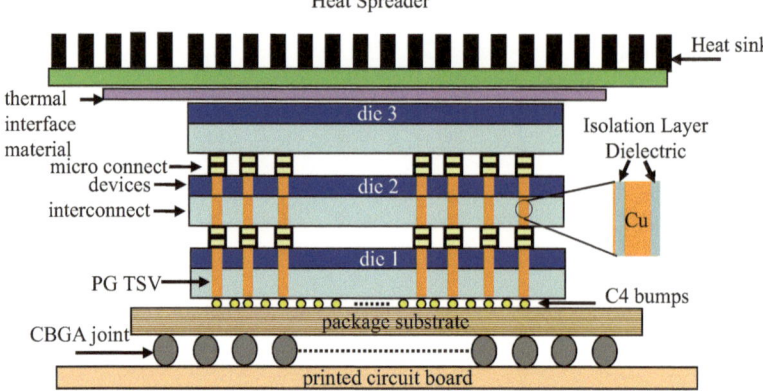

Fig. 2.1 A 3-D IC, illustrating various components of a 3D system

Fig. 2.2 A regular
cylindrical TSV

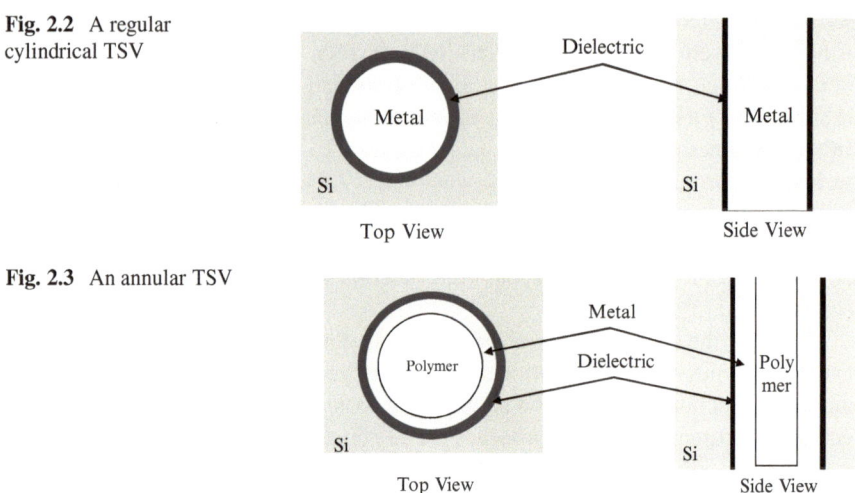

Fig. 2.3 An annular TSV

model than other TSV types. Substrate noise and thermo-mechanical stress are the main challenges. The cylindrical via shape allows for a more uniform insulation layer, and hence a higher breakdown voltage, than for the square-shaped via [101].

2.1.2 Annular TSV

An annular TSV, shown in Fig. 2.3, consists of a polymer-filled core surrounded by a metallic (usually Cu) annulus, which is separated from substrate by a dielectric (usually SiO_2) layer. This TSV type was proposed to overcome manufacturing challenges of cylindrical TSVs [104]. An annular TSV can be fabricated using a

Fig. 2.4 A tapered TSV

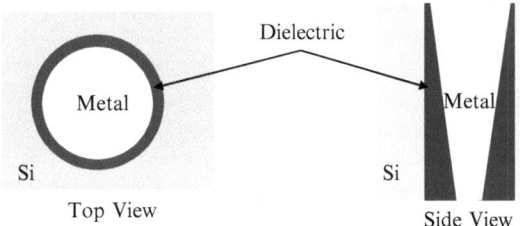

simple manufacturing process at a lower cost compared to regular TSVs. IBM has developed two CMOS compatible processes to fabricate annular TSVs [8, 9]. To achieve the same conductivity, the required cross sectional area is smaller for an annular TSV than for a regular TSV. This reduced area results in improved thermo-mechanical stability. Xie et al. provide a detailed thermo-mechanical analysis of annular TSVs [116].

2.1.3 Tapered TSV

The tapered TSV, shown in Fig. 2.4, was introduced by MIT Lincoln Laboratory and has been built in SOI technology [23]. Tapered TSVs in SOI technology do not require insulation and have lower capacitance than regular TSVs. Tapered TSVs are preferred over regular TSVs because of their simple fabrication process [55] but tapering results in a more resistive TSV when compared to a non-tapered one with the same maximal cross-sectional area. Excessive tapering can be problematic as it leads to V-shaped vias. A specific interconnect pitch is required at the bottom side of the TSV. A process technology to precisely control the via slope is demonstrated by Nagarajan et al. [80].

2.1.4 Coaxial TSV

Thin dielectric liner around a TSV and its large extension through the substrate can create electrical coupling and critical substrate noise in neighboring active devices and TSVs [53,90]. A coaxial TSV, first proposed by Sparks et al. [97] and illustrated in Fig. 2.5, is a regular TSV with an added surrounding metal layer. The inner metal layer can be used for signal transmission while the outer metal layer, connected to circuit ground, provides shielding. Although a few processes have been proposed to fabricate coaxial TSVs [35, 44, 45], no mature technology currently exists to fabricate them.

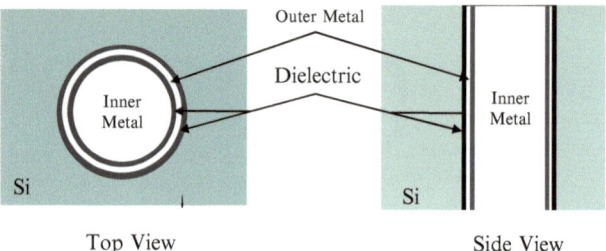

        Top View   Side View

Fig. 2.5 A coaxial TSV

2.2 TSV Integration

Two common TSV integration techniques are shown in Fig. 2.6. Bonding of dies using controlled collapse chip connect (C4) bumps and TSVs is shown in Fig. 2.6(a). C4-based TSV integration is the simplest among the two techniques because C4 bumps is a mature and well-understood technology. At the same time, large C4 dimensions are the bottleneck to achieve dense inter-die interconnects and reduced package height.

Thin film bonding, shown in Fig. 2.6(b), eliminates the C4 bumps offering increased interconnect density and wafer-scale bonding. The two common bonding processes are dielectric and metallic bonding. In dielectric (oxide or polymer based) bonding, vertical connections are completed after the bonding process. This bonding technique uses through-strata or 3-D vias that pass through the top die and connect to the conventional interconnect in the bottom die [2, 106]. In metallic bonding, the vertical connections are formed by bonding conductive microconnects of Cu, or Cu with a plating of Tin, on each bonding surface [7, 76]. The bonded microconnects typically have a pitch in the range of 20–60 μm, which is expected to improve in the future to allow for higher-density inter-layer connections.

Figure 2.7 shows two topological arrangements to stack dies: front-to-back (F2B) and front-to-front (F2F), where "front" refers to the metal interconnect side and "back" refers to the Si substrate side. F2F bonding does not require TSVs for inter-die communication and results in higher speed inter-die communication than the ones that use F2B bonding. Power delivery and all other interconnect utilize TSVs. For F2B bonding, die thinning is required to reduce TSV height [66].

2.3 TSV Fabrication

TSV fabrication is a vital component of the 3-D IC design process. Key steps include via etching, sidewall insulation, via filling, wafer thinning, and wafer/die stacking [17,18,71]. Quality and maturity of these process steps determine TSV performance.

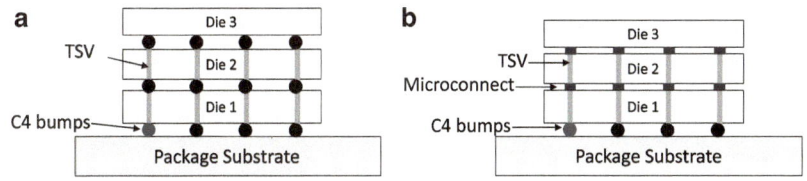

Fig. 2.6 TSV integration techniques. (**a**) TSV integration with C4 bonding, (**b**) TSV integration with metallic/dielectric bonding

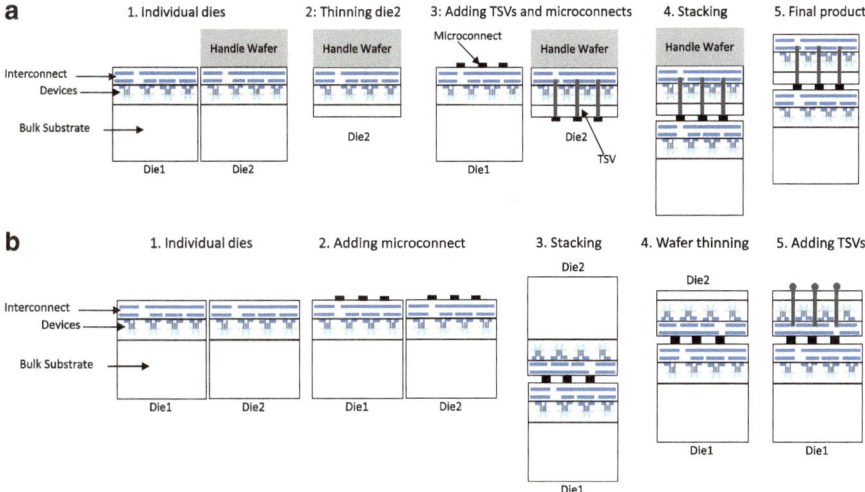

Fig. 2.7 Bonding processes for different topological options for 3-D stacking. (**a**) Front-to-back bonding, (**b**) Front-to-front bonding

IC fabrication can be divided into two major stages: front-end-of-line (FEOL) and back-end-of-line (BEOL). All processing steps preceding the first metal layer are part of FEOL whereas the rest of the processing steps lie in BEOL. Based upon this distinction, TSV fabrication processes can be divided into three categories. These categories are described below and summarized in Table 2.1.

2.3.1 Via First

This is the simplest process where TSVs are fabricated in the bulk substrate before any devices or interconnect layers are formed [61]. This process does not require any backside lithography. Moreover, any of the FEOL-compatible processes, including high temperature thermal oxidation, can be used to deposit TSV insulation. To make this process FEOL-compatible, the TSV conducting material must be polySi.

Table 2.1 Comparison of via formation options

	Via first	Via middle	Via last
TSV manufacturing source	Foundry, IDM	Foundry, IDM	Foundry, IDM, OSAT
Etched materials	Silicon	Oxide, Silicon	Oxide, Silicon
TSV fill material	Poly-silicon	Copper, Tungsten	Copper
TSV fill method	CVD	CVD, ED Cu	Sputter, ED Cu
Temperature restrictions	None	$<450^\circ$C	$<450^\circ$C
Applications	Memory	Generic	Image sensors

Because TSVs are fabricated in the front-end process, in a foundry, highly reliable (electrically as well as mechanically) TSVs can be formed with densities larger than 100,000 per cm^2 and aspect ratios of more than 25:1 [110].

2.3.2   Via Middle

In this process, TSVs are created after the device fabrication but before any interconnect layers are formed [9, 89]. TSV conducting material may include Cu or tungsten (W) deposited by chemical vapor deposition (CVD) or electric discharge (ED) [110].

2.3.3   Via Last

In the via last process, TSVs can be fabricated anytime during and after BEOL [9, 89]. Typically, backside lithography is used to fabricate these vias. For this purpose, a handle-wafer is attached to the front side of the wafer to be processed. The wafer is thinned and vias are formed from the backside of substrate. This process is also shown in Fig. 2.7(a). These TSVs need to be connected to lower metal layers (M1 or M2) of the die to avoid etching through the higher metal layers. This process requires a low-temperature TSV insulation technique to minimize thermal effects on already fabricated devices and interconnect. Other challenges include backside lithography, wafer-thinning and handling, and reliably opening the insulation at the base of the vias to make reliable electrical connections.

2.4   Electrical Modeling of TSVs

Using resistive, inductive, and capacitive elements, a TSV can be modeled as a circuit component. TSV performance (signal delay and power dissipation) is governed by the values of these RLC elements. Various approaches have been considered to extract relevant RLC parameters. We describe an overview of these approaches.

The simplest approach is to estimate the R and C values of a TSV using basic calculations:

$$R = \frac{\rho_m h_{TSV}}{A_{eff}} \tag{2.1}$$

$$C = \frac{\varepsilon_o \varepsilon_r S_a}{t_{ILD}} \tag{2.2}$$

where h_{TSV} is the TSV height, A_{eff} is the TSV cross-section area, S_a is the sidewall area, t_{ILD} is the dielectric liner thickness, ρ_m is the resistivity of TSV material, and ε_r and ε_o are the relative permittivity of SiO_2 and permittivity of free space, respectively. Alam et al. calculate TSV resistance and capacitance using this analytical model as well as using three-dimensional electrostatic simulations [7]. For a 50 μm high square TSV with a 5 μm side and a 1 μm sidewall dielectric thickness, the capacitance and resistance values are 40 fF and 43 mΩ, respectively.

Dong et al., on the other hand, fabricate a ground-signal-ground TSV circuit [49]. S-parameters for this circuit are measured using micro-probing. A schematic drawing of an equivalent circuit model is shown in Fig. 2.8. The circuit parameters are determined by fitting the S-parameters from the model to the measured S-parameters from the fabricated TSVs. Table 2.2 shows the values of the circuit parameters for a TSV of diameter 55 μm and pitch 150 μm. R_{via0} and L_{via0} are the values of resistance and inductance at 0.1 GHz and their values are assumed to be 12 mΩ and 35 pH, respectively.

Sun et al. fabricate different types of blind vias to study the electrical characteristics of dielectric liner [101]. Square and cylindrical TSVs of height 80 μm and thickness in the range of 20–100 μm are fabricated. Breakdown voltage and liner capacitance are measured to be 40 V and 50–60 fF, respectively.

Finally, accurate inductance characterization is more complex as it is essential to include a return path directly based on the design and layout of specific interface circuitry. Multiple studies based on simulation as well as test structure measurements demonstrate that TSV inductance is low, in the range of 0.3–0.9 pH per μm of TSV length [7, 88, 114].

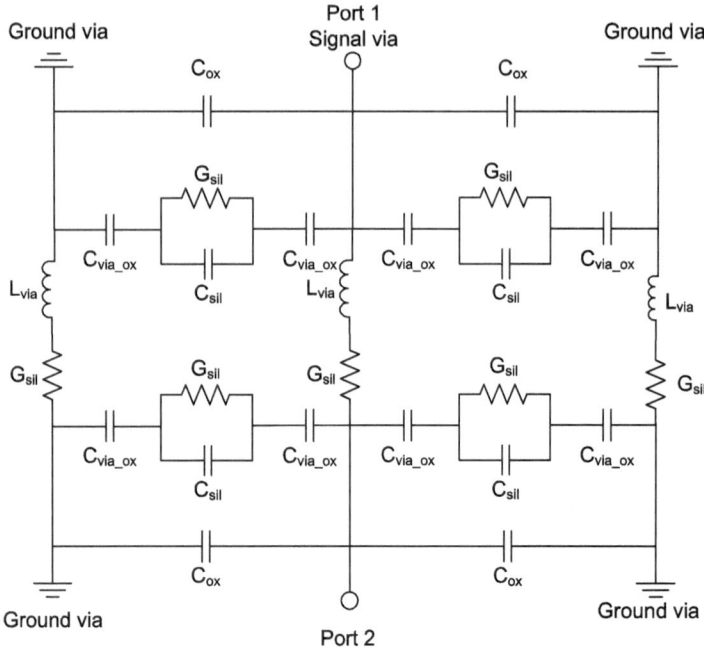

Fig. 2.8 Electrical circuit model of TSV

Table 2.2 Circuit parameters for TSV of diameter 55 μm

Parameter	Description	Value
R_{via}	TSV resistance	$R_{via0}\sqrt{1+\frac{f}{10^8}}$
L_{via}	TSV inductance	$\dfrac{L_{via0}}{1+log(\frac{f}{10^8})^{0.26}}$
C_{via-ox}	Capacitance of TSV liner	910 fF
C_{ox}	Capacitance of the oxide layer on the silicon surface and the fringing field between the vias	3 fF
C_{sil}	Capacitance of the silicon substrate	9 fF
G_{sil}	loss property of the silicon substrate between the signal via and the ground via	1.69 m/Ω

2.5   Summary

We presented in this chapter a summary of existing TSV types (regular, annular, tapered, and coaxial) and provided an overview of bonding and integration techniques. In this book, we use regular TSVs when analyzing TSV-induced substrate noise, minimizing TSV area dedicated for power delivery, and when utilizing carbon nanotubes. Both regular and coaxial TSVs are explored when designing PDNs in Chap. 4. The presented work however can be extended to deal with all TSV types.

Chapter 3
Analysis and Mitigation of TSV-Induced Substrate Noise

TSVs are a major source of substrate noise that threatens the performance of neighboring devices. In addition, TSV noise increases leakage current, which increases static power consumption and can erroneously switch transistors off or on [91]. A "keep out" zone, specified through layout rules, is thus required to shield devices from neighboring TSVs.

We study in this chapter the TSV-induced substrate noise problem and analyze the effectiveness of different noise mitigation techniques. We propose a practical and effective device, a ground (GND) plug, to reduce TSV-induced substrate noise. A GND plug is a TSV-like structure that is connected to circuit ground and may extend partially or completely through the substrate. Multiple GND plugs, fabricated around a TSV, provide noise isolation between TSVs and neighboring devices. We examine the physical design and placement of GND plugs for effective noise isolation. We compare the GND plug technique with two other noise mitigation techniques: thicker dielectric liner and backside ground plane. Thicker dielectric liner provides shielding that decreases coupling between TSV and substrate. A backside plane, electrically connected to circuit GND, ensures sufficient substrate grounding.

The rest of the chapter is organized as follows. We present an overview of TSV-induced noise and related work in Sect. 3.1. We describe an evaluation framework used to perform lumped parasitic analysis in Sect. 3.2. We perform a comparison of three different noise mitigation techniques: thicker dielectric liner, backside ground, and GND plugs in Sect. 3.3. We provide the summary in Sect. 3.4.

3.1 Problem: TSV-Induced Noise

Figure 3.1 shows a cross-section view of a Silicon (Si) substrate with a TSV and a MOSFET transistor. Signal transitions through a TSV create noise that can pass through the substrate and change the body voltage (V_B) of neighboring

N. Khan and S. Hassoun, *Designing TSVs for 3D Integrated Circuits*, SpringerBriefs in Electrical and Computer Engineering, DOI 10.1007/978-1-4614-5508-0_3,

Fig. 3.1 Cross-section view
illustrating TSV-to-device
coupling

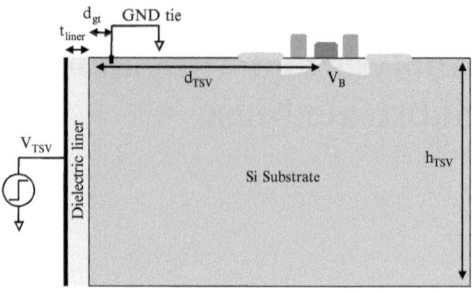

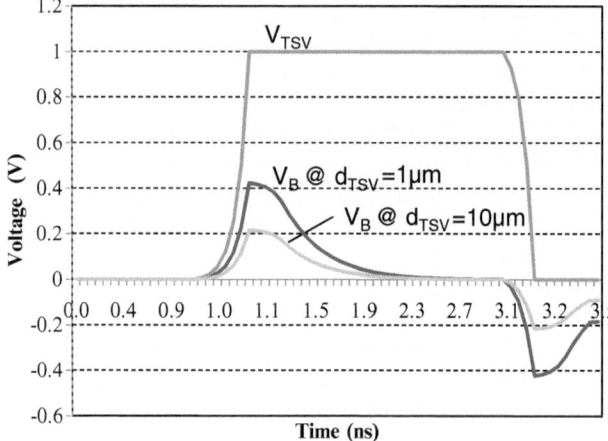

Fig. 3.2 Body voltage during TSV signal transition at different TSV distances, d_{TSV}, for $V_{TSV} = 1$ V square wave, $h_{TSV} = 20\,\mu m$, $t_{liner} = 1\,\mu m$, $d_{gt} = 0.5\,\mu m$, signal transition time = 50 ps [53]

transistors. This change in V_B impacts the performance of devices. Physical design considerations that can be exploited to mitigate TSV-induced substrate noise include dielectric liner thickness (t_{liner}), TSV-to-device distance (d_{TSV}), and GND ties (conventional substrate ties, often referred to as substrate contacts). TSV-to-device coupling can be controlled by t_{liner} and d_{TSV} but adding more GND ties provides sufficient grounding of the substrate.

Figure 3.2 shows variations in the device body voltage, V_B, at different distances from a TSV for a set of design parameters. These transitions are short-lived and occur with a change in the signal passing through the TSV. For a 1 μm thick liner, the peak value of these transitions is significant (40% of VDD), despite including a GND tie 0.5 μm from the TSV.

To explore the impact of body voltage variations on device performance, we modeled a fan-out of 4 inverter in 32 nm technology node. We vary the peak body voltage, keeping the waveform the same as shown in Fig. 3.2. The resulting variations in delay and dynamic power are shown in Fig. 3.3. For a peak body

Fig. 3.3 Variations in performance and power of a fan-out-of-four CMOS inverter in 32 nm technology node due to body voltage noise

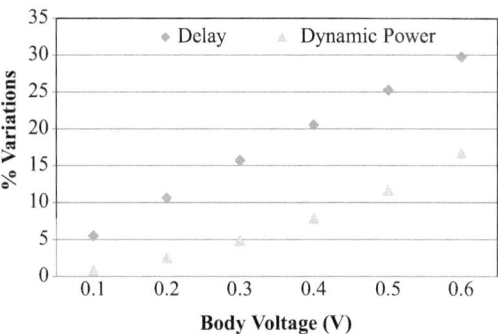

voltage of 0.4 V, delay is increased by 21% while dynamic power is increased by 8%. It is evident from Figs. 3.2 and 3.3 that TSV-induced noise plays a significant role in determining device performance.

Substrate noise is a well-studied problem in traditional 2-D IC design [6]. While a manageable concern in digital circuit design, analog circuit design has always been subject of higher scrutiny for noise isolation. Several noise isolation techniques, including split power planes, deep-nwell process, and guard ring structure, have been employed in mixed-signal designs. For a 3-D IC design, this problem is not yet well-addressed. The extent of the TSV-induced substrate noise problem is directly related to the density of TSVs. Conventional noise mitigation techniques, such as isolated floorplanning of noise sensitive circuits with guard bands, are not feasible in 3-D designs with the high TSV density as predicted by the ITRS roadmap [5].

A number of techniques have been identified for noise mitigation: thicker dielectric liner, backside ground plane [53], guard ring structure [30], and co-axial TSVs [45, 53]. Increasing the thickness of the dielectric liner surrounding a TSV is the simplest approach. Increased liner thickness has already been shown to be insufficient in mitigating substrate noise [53]. Providing a backside ground by placing a die on a grounded metal sheet is a common strategy to mitigate substrate noise in 2-D ICs. This strategy may not be practical for 3-D ICs because of two reasons: (1) a metallic sheet between dies will introduce unnecessary inductive coupling, (2) design complexity will increase because TSVs passing through metallic sheets must be isolated. Surrounding TSVs with guard rings is not effective because typical guard-ring depth is comparable to GND tie depth, which is too small to provide any significant isolation [30]. Lastly, using a co-axial TSV is promising to mitigate noise but the manufacturability of co-axial TSVs is still in question.

We investigate in this chapter an alternative and more practical technique to reduce TSV-induced substrate noise. We use tungsten (W)-filled GND plugs in the vicinity of TSVs. W-filled TSVs have been demonstrated for the fabrication of 3-D LSI chips [54]. Unlike TSVs, the proposed W-filled GND plugs may not extend the complete depth of the substrate. Moreover, GND plugs are filled with W which has higher resistivity than Cu, the preferred TSV fill. W-filled plugs have the advantages of direct substrate connectivity without any barrier layer and higher RC damping for less noise transfer from the GND power supply. To quantify the effectiveness of the proposed W-filled GND plugs and to investigate relevant physical design parameters, such as placement and height, we first develop a lumped parasitic analysis framework.

3.2 Evaluation Framework for TSV-Induced Noise

We use a three-dimensional evaluation framework composed of a Cu TSV in an Si substrate, ground ties, and voltage observation points. The top view of this setup is shown in Fig. 3.4. Each component is defined below.

- **Substrate:** We assume a high-R substrate with a resistivity of 10Ω-cm and relative permittivity of 11.8. This type of substrate is used to fabricate low cost, low performance devices like memory [31]. We assume a $50\times50\,\mu m$ substrate. This cross-section is sufficiently larger than the TSV and device scale for capturing their interactions without interference from other devices or structures in a die. The boundary condition at the sidewalls of the $50\times50\,\mu m$ substrate is such that we have zero-current going out of the surfaces. The condition is in synergy with (a) multiple $50\times50\,\mu m$ modeled substrate neighboring each other form the die and (b) there is no interaction/interference between them.
- **TSV:** We assume a cylindrical Cu TSV. Its height is the same as the substrate height and the diameter is fixed to $2\,\mu m$.
- **Dielectric Liner:** We use an SiO_2-based liner, with $10^{16}\Omega$-cm resistivity and 3.9 relative permittivity, surrounding the TSV. The default thickness of the dielectric liner is assumed to be $0.1\,\mu m$, which is consistent with recent design studies [85].
- **Shallow Trench:** Thermal stress is one of the most important factors shaping TSV technology. We assume an SiO_2-based shallow trench. Its thickness and depth (into the top surface of substrate) are assumed to be $0.90\,\mu m$ and $0.3\,\mu m$ respectively. These values are consistent with those in the ITRS roadmap [5].
- **Observation Points:** We assume nine equally spaced observation points (P1, P2, ...) located $4–20\,\mu m$ away from the center of the TSV. These points are modeled as small metallic cubes to enable extracting parasitics between TSV and devices at various distances from the TSV.

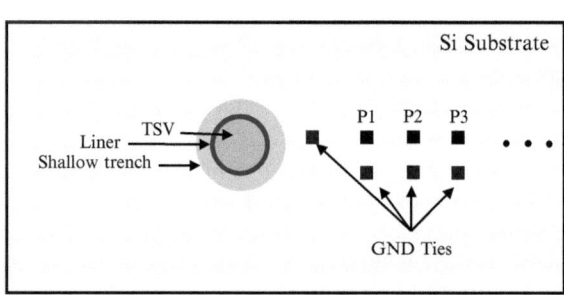

Fig. 3.4 Top view of TSV-induced noise analysis framework

Table 3.1 Critical parameters for our TSV-induced noise analysis framework

Parameter	Value
Substrate height	20 μm
Substrate length	50 μm
Substrate width	50 μm
TSV height	20 μm
TSV diameter	2 μm
Liner thickness	0.1 μm
Shallow trench height	0.3 μm
Shallow trench width	0.9 μm
Resistivity of high-R substrate	10 Ω-cm

Fig. 3.5 An example extracted circuit comprising TSV parasitics, a single observation point, a GND tie, and Si substrate parasitics for coupling among the three elements

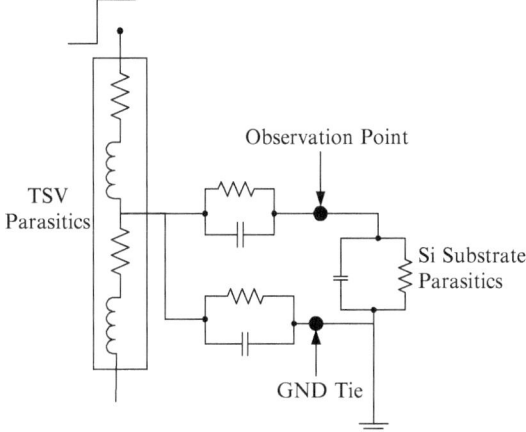

- **GND Ties:** Placing GND ties throughout the circuit layout is the conventional approach to control transistor body voltages, and hence, is considered in our setup. We assume a GND tie at a distance of 0.3 μm from the shallow trench edge. Also, we assume that there is at least one GND tie within a 1 μm distance of each observation point. This GND tie is not the proposed W-filled GND plug.

Table 3.1 shows the default values of parameters in our TSV-induced noise analysis framework. These values should be assumed when any of the parameters is not being varied for sensitivity investigation. To extract an equivalent SPICE circuit for our framework, we use a finite-element based 3D extraction tool (Q3D Extractor) from Ansoft. Figure 3.5 shows a portion of the extracted circuit comprising a TSV, an observation point, a GND tie, and Si substrate. A step input, with a rise time of 100 ps and peak voltage of 1 V, is applied at one of the TSV terminals whereas the other terminal is assumed to be floating. We perform transient analysis and report peak noise at the observation points.

RLC values of the extracted circuit depend upon the extraction frequency. Q3D Extractor divides the frequency spectrum into DC, AC, and transition regions. In contrast to the AC region, the DC region does not account for any skin depth and relative frequency dependent parameters. The transition region, which spans about a decade of frequency, does not produce a valid solution because neither AC nor DC models are truly valid [1]. Before we perform any analysis, we extract the SPICE netlist for various values of operating frequencies (between 1 KHz and 10 GHz) using the default design parameters from Table 3.1. Due to the relatively large dimensions of TSVs, on the order of microns, the DC region extends well above 1 GHz frequency. We did not observe any changes in the RLC parameters in this frequency range. We choose the operating frequency to be 1 GHz for our analysis.

3.3 TSV-Induced Substrate Noise Analysis

We present in this section an analysis of different noise mitigation techniques. We investigate the impact of using thicker dielectric liner and using a backside ground plane in Sect. 3.3.1 and in Sect. 3.3.2 respectively. We present a detailed study of the proposed GND plug technology in Sect. 3.3.3. We compare the performance and area penalty of these three techniques in Sect. 3.3.4.

3.3.1 Thicker Dielectric Liner

Increasing liner thickness is the simplest approach to mitigate TSV-induced substrate noise. We vary liner thickness from 0.1 to 3 μm in the default setup shown in Fig. 3.4 and extract an RLC circuit for each setup using Q3D Extractor. SPICE is used for simulation. We record peak transient noise at various observation points in the substrate. Figure 3.6 plots peak transient noise for several liner thickness values at observation distances 6, 12, and 18 μm from the TSV. We conclude the following:

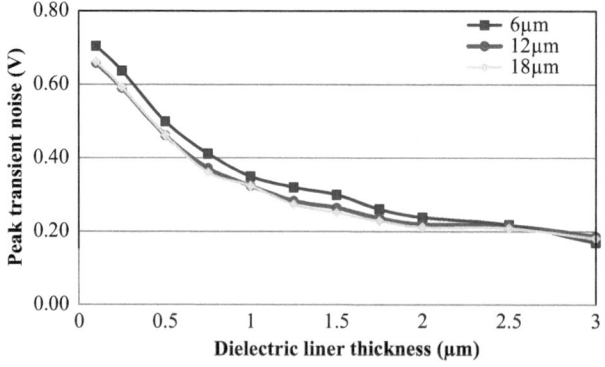

Fig. 3.6 Impact of thicker liner: peak transient noise at different observation distances (6, 12, and 18 μm) from the TSV for increasing liner thickness

- Peak transient noise ranges from 0.18 to 0.7 V, regardless of distance from TSV, for all examined values of liner thickness. This indicates that standard GND substrate ties are inadequate for creating a reference GND substrate in the presence of a TSV switching signal. Hence, using GND substrate ties alone is not effective in mitigating TSV-induced noise.
- Peak substrate noise decreases with increasing liner thickness. This trend is not uniform and can be divided into three segment. The impact of increasing liner thickness is maximal for liner thickness between 0.1 and 1 μm, reduces for liner thickness between 1 and 2 μm, and saturates for thickness greater than 2 μm.
- Peak substrate noise is ≈18% of VDD for liner thickness of 3 μm. For 2 μm diameter TSV with t_{liner} 0.1 μm and t_{ST} 0.9 μm, increasing t_{liner} by 2.9 μm results in a 6× area penalty ($12.6 \mu m^2$ vs. $75.4 \mu m^2$). This huge area penalty, creates large interconnect blockages and reduces the area available for active devices.

3.3.2 Backside Ground Plane

During assembly and packaging stages, a 2-D die is placed on a grounded metal layer. As mentioned in Sect. 3.1, the same idea can be extended to 3-D ICs where the substrate has a backside grounded metal, in preferable plate or grid format, creating a strong GND reference for substrate. To model this technique, we add a Cu sheet in the default setup shown in Fig. 3.4. The sheet cross-section is the same as the substrate cross-section. The sheet thickness is assumed to be 2 μm. One side of the sheet connects to the substrate and the other side connects to GND. Substrate thickness, the distance between the devices layer and the backside ground, is the only variable of concern for substrate noise analysis. We vary the substrate thickness from 10 to 40 μm and extract the RLC circuit for each setup. We use SPICE to simulate and record peak transient noise at the observation points. The results of this study are shown in Fig. 3.7. We make the following observations:

- Substrate noise decreases as a function of TSV distance. Using the backside ground plane is more effective than the thicker dielectric in localizing noise.
- Our results show that the capability of the backside ground to mitigate substrate noise is a function of substrate thickness. The benefits are minimal for larger values of thickness. Backside ground is therefore effective in technology generations where substrate heights can be aggressively thinned.

3.3.3 GND Plugs

The analyses presented above show that both noise mitigation techniques have their performance and/or technology limitations. As an alternative technique, we propose using GND plugs. We propose Tungsten (W) as the fill material for two reasons.

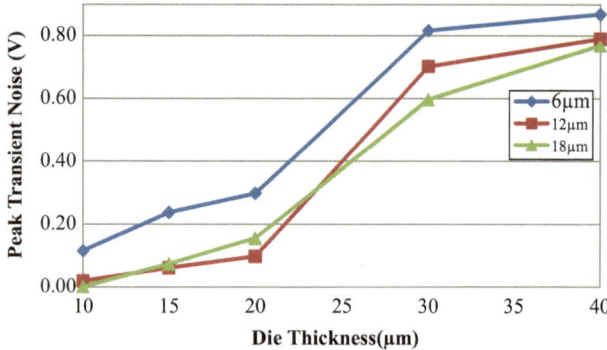

Fig. 3.7 Impact of a backside ground plane: peak transient noise at different observation distances (6, 12, and 18 μm) from the TSV for decreasing die thickness

Fig. 3.8 Top view of
TSV-induced noise analysis
framework with GND plugs

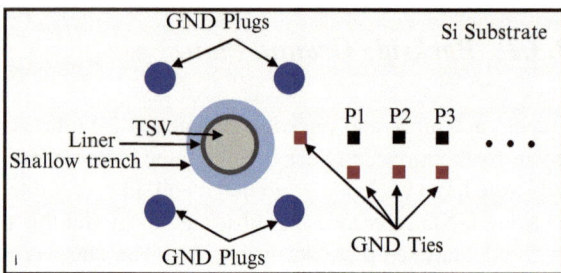

First, a smaller CTE (Coefficient of Thermal Expansion) mismatch of W and Si, compared to Cu and Si, will result in less thermal stress in devices. Second, unlike Cu, W does not require a diffusion barrier and will provide a direct connection between substrate and circuit ground resulting in better device shielding. A GND plug is different from a GND substrate tie that provides typical substrate or well ties without much depth into the substrate.

Figure 3.8 shows the design setup with added GND plugs. We first investigate the impact of placing a different number of GND plugs. We evaluate the use of four GND plugs fabricated at 3 μm from the center of the TSV. Noise isolation is improved by 2.46× when compared to using only two GND plugs. For the rest of the analysis, we therefore utilize four plugs.

Next, we explore the impact of two critical parameters: plug diameter and plug depth. RLC circuits for each setting are extracted using the Q3D Extractor and then simulated in SPICE for peak transient voltage at the observation points. Figure 3.9 shows the peak transient noise for three different GND plug diameters with varying plug height. We make the following observations:

- A GND plug is effective in reducing peak noise. A deeper GND plug is more effective than a shallower one.
- The ability to localize noise increases with an increase in GND-plug depth. GND plugs with depth greater than 75% of the substrate height are required

Fig. 3.9 Impact of GND plug height on peak transient noise at different observation distances (6, 12, and 18 μm), (**a**) Plug diameter = 0.5 μm, (**b**) Plug diameter = 1 μm, (**c**) Plug diameter = 1.5 μm

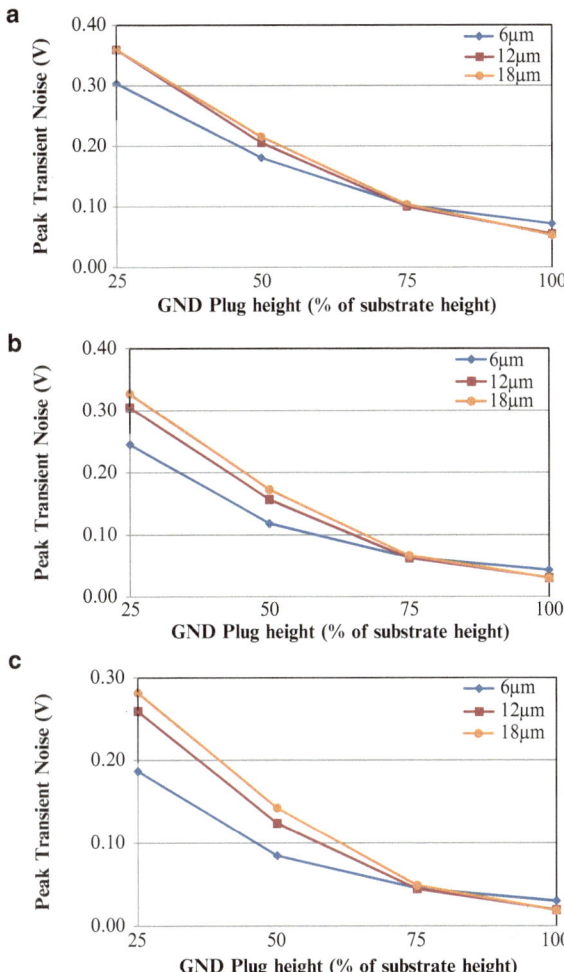

to localize noise. A shallow GND plug provides only partial shielding; devices farther away from the TSV will not be as thoroughly shielded as with a GND plug that completely extends through the substrate.

- The benefit of increasing plug diameter occurs at the expense of increased area. Increasing the plug diameter for deep GND plugs has larger relative impact than using the same diameter for shallow GND plugs. For a 3× increase in plug diameter from 0.5 to 1.5 μm, the average noise reduction is 2×.

GND plug diameter and depth are related due to aspect ratio limitations of deep trench formation and filling in Si substrate. The results in Fig. 3.9 suggest the need

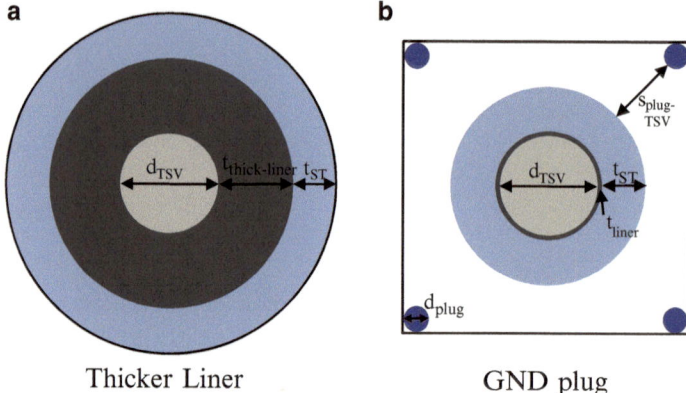

Fig. 3.10 Comparison of substrate area dedicated for thicker liner and GND plugs (d_{TSV} =2 μm, $t_{thick-liner}$ =1.5 μm, t_{ST} =0.9 μm, t_{liner} =0.1 μm, $s_{plug-TSV}$ =1 μm, and d_{plug} =0.5 μm)

for small diameter yet deep GND plugs. Kikuchi et al. demonstrate W-filled TSV formation using deep-Si-trench etching and Tungsten chemical vapor deposition (CVD) where TSV diameter is maintained till 70% of substrate height for an aspect ratio of approximately 18:1 [54]. Interestingly, the proposed GND plug scheme does not require sidewall isolation nor high uniformity of plug diameter as a function of depth. A higher aspect ratio cone-shaped plug is thus possible, and warrants further investigation.

3.3.4 Comparison of Three Noise Mitigation Techniques

In addition to peak noise, area penalty and routing blockages are important comparison metrics of the three discussed noise mitigation techniques. We utilize the default design parameters described in Table 3.1 and explore the relative merits of the three techniques. In the thicker dielectric liner approach, we assume the liner thickness to be 1.5 μm. In the GND plugs approach, we consider two cases: (1) 0.5 μm diameter plugs with the same depth as the substrate thickness, and (2) 0.5 μm diameter plugs with a depth equal to 50% the substrate thickness. The top views of a TSV with thicker liner and a TSV with GND plugs are shown in Fig. 3.10 to illustrate area penalty analysis.

Table 3.2 reports peak transient noise and substrate area blockages for the different noise mitigation techniques along with the baseline TSV case where no noise isolation technique is applied. The peak noise is reported at 6, 12, and 18 μm away from the TSV center. The proposed GND plug technique with a 40:1 aspect ratio is superior in mitigating the TSV-induced substrate noise by an order of

Table 3.2 Comparison of peak transient noise and relative area blockage for the three TSV-induced noise mitigation techniques

Technology	Peak transient noise (V) at distances:			Substrate area blockage (relative)
	6 μm	12 μm	18 μm	
Baseline TSV (TSV height = 20 μm, liner thickness = 0.1 μm)	0.705	0.658	0.663	1
Thicker liner (1.5 μm)	0.300	0.265	0.251	3.06
Backside ground plane for baseline TSV	0.298	0.098	0.155	1
GND plug (diameter = 0.5 μm, height=20 μm) with baseline TSV	0.072	0.055	0.053	2.0
GND plug (diameter = 0.5 μm, height=10 μm) with baseline TSV	0.181	0.206	0.216	2.0

magnitude when compared to using thicker liner and backside ground approaches. For the design parameters, shown in Fig. 3.10, the area penalty for the thicker liner is $36.3\,\mu m^2$ and for the GND plug is $24.5\,\mu m^2$.

Fabrication of W-filled TSV with an aspect ratio of 50:1 and diameter of 1 μm has been proposed [77], which suggests that the GND plug with a diameter of 0.5 μm and an aspect ratio of 40:1 is achievable. We present the results for a smaller aspect ratio of 20:1 in Table 3.2. These results show that even the smaller aspect ratio yields 40% reduction in TSV-induced noise when compared to the other two approaches. If the desired substrate noise limit is 10% of VDD, then the GND plug with a height of 20 μm is the only solution that safely enables placing a device 6 μm from the TSV center. The resulting keep out zone considering TSV-induced noise is thus a square area of side 5 μm. When using a thicker liner, a similar noise margin can only be attained using a liner thickness greater than 3 μm, resulting in a keep out zone diameter in excess of 10 μm. TSV-induced stress also contributes to the required keep out zone but it is not considered in our analysis.

3.4 Summary

We studied in this chapter the problem of TSV-induced substrate noise. We proposed a novel noise mitigation technique, a GND plug, and compared its effectiveness against two other noise mitigation techniques: thicker dielectric liner and backside ground plane. We assumed practical design parameters and utilized a three-dimensional finite-element based solver to extract the SPICE netlist of our experimental setup. Analysis of a 1.5 micron-thick dielectric liner shows peak substrate noise of 30% of VDD, thus necessitating further increase in thickness or a significant increase in the keep out zone. Furthermore, the resulting area penalty, 3× the size of a 2 μm diameter TSV, creates routing blockages and reduces the area available for active devices. While a backside ground plane or mesh is effective with thinned dies, placing such a metal sheet or mesh between dies in 3-D ICs may not be practical. We showed that a GND plug with an aspect ratio of 40:1 is effective in reducing noise by an order of magnitude with a smaller area penalty than a thicker liner. A GND plug with a smaller aspect ratio (20:1) provides 40% reduction in substrate noise when compared to the other two techniques. The proposed GND plug technique thus offers a practical and promising solution to the difficult problem of providing device shielding against TSV-induced substrate noise.

Chapter 4
TSVs for Power Delivery

Robust power delivery is one of the ITRS scaling grand challenges due to increasing operating frequencies, increasing power density, and decreasing supply voltages. Three dimensional stacking of multiple dies makes this problem even more challenging. In a 3-D IC, only the die adjacent to the package can get power directly from the package. Dies away from the package require new technologies for power delivery. We evaluate in this chapter using TSVs to deliver power in a 3-D IC with the goal of understanding factors that contribute to the performance of a 3-D power delivery network (PDN). We investigate the impact of TSV size. We study various architectural configurations to find the best TSV granularity. We explore the impact of shared and dedicated TSVs on PDN performance and the feasibility of coaxial TSVs for power delivery.

We measure PDN performance in terms of maximum, average, and standard deviations of IR drops and Ldi/dt droop. These metrics quantify local and global PDN characteristics in both dc and transient analysis. Our evaluation framework consists of a four-core chip multiprocessor (CMP), a memory (MEM), and an accelerator engine (ACCL). We use realistic workloads from SPEC benchmarks for each functional module in the system. The PDN for our 3-D designs is modeled using both off-chip and on-chip components.

The rest of the chapter is organized as follows: Sect. 4.1 provides an overview of power delivery in 3-D ICs and related work. Section 4.2 describes our design setup. TSV size is optimized in Sect. 4.3. Different comparative studies to find the best TSV granularity are presented in Sect. 4.4. We study the impact of dedicated TSVs in Sect. 4.5. Analysis of power delivery using coaxial TSVs is provided in Sect. 4.6. Design guidelines are presented in Sect. 4.7. The work is summarized in Sect. 4.8.

N. Khan and S. Hassoun, *Designing TSVs for 3D Integrated Circuits*, SpringerBriefs in Electrical and Computer Engineering, DOI 10.1007/978-1-4614-5508-0_4, © The Authors 2013

4.1   Problem: Power Delivery for 3-D ICs

3-D integration poses grand power delivery challenges for two reasons: increased power density and package asymmetry. Contrast a 3-D IC with a functionally comparable 2-D IC. The average wire length for a 3-D IC drops by a factor of $N^{1/2}$ where N is the number of dies in the 3-D IC, and the wire resistance and capacitance decreases proportionally [34]. Assuming that the design is interconnect-dominated, power is expected to drop by a factor of $N^{1/2}$. If the power density of each die in the 3-D IC is similar to that in the 2-D case and each die size is $1/N^{th}$ of that in the 2-D case, the power density per square area for the stacked 3-D chip increases by a factor of $N^{1/2}$. The power delivery requirements thus increase with the number of dies in the stack.

To understand the impact of package asymmetry on power delivery, consider the illustrative example in Fig. 4.1. Three dies are stacked between the heat sink and the package substrate. Electric signals and power are routed from the printed circuit board to the package substrate through ceramic ball grid array (CBGA) joints, and then they are distributed utilizing controlled collapsed chip connections (C4). The dies are bonded using microconnects. Through-silicon vias (TSVs) pass through a die and provide electrical connectivity for signals or power delivery among the dies. Clearly, the package asymmetry impacts both power delivery and heat removal, another critical challenge in 3-D ICs. While thermal issues have received considerable attention, e.g. [32, 41, 56, 119], 3-D power delivery has not yet been adequately addressed.

Previous work on 3-D power delivery can be summarized under two main themes: power delivery techniques and power integrity analysis. Kim et al. analyzed a multistory power delivery technique where a higher than nominal VDD supply voltage is applied from the package and distributed differentially to subsequent power rails using level conversion [48]. Their work utilized lumped off-chip and on-chip models with tungsten filled TSVs in bonded SOI technology to

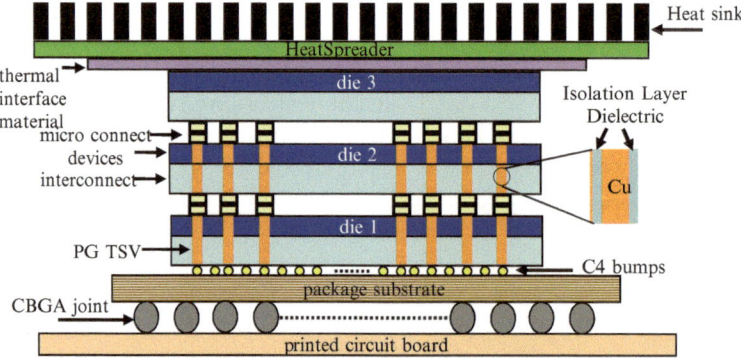

Fig. 4.1 Illustrative 3D system assuming face-to-back metallic bonding with micro-connects

assess the impedance response of overly simplified lumped 2-D and 3-D PDNs. Yu et al. investigated the impact of via stapling, where a 3-D mesh is created, on both power and thermal integrity [119]. Zhan et al. proposed a partition-based algorithm for assigning modules at the floorplanning level to reuse currents between VDD domains, and to minimize power wasted during circuit operation [121]. In the power integrity analysis area, Huang et al. proposed an analytical physical model of 3-D PDN, accurate within 4% compared to SPICE, to capture the impact of power supply noise [46]. The allocation of decoupling capacitors, in a 3-D IC, has also been investigated [72, 113, 123].

Most of the previous works assume worst case switching currents and utilize overly simplified power grid network models. In contrast, our work utilizes a more detailed off-chip and on-chip PDN in a realistic design example where we use a workload derived from SPEC benchmarks. We estimate both IR drop and Ldi/dt droop in 3-D PDN. In addition to quantify the impact of TSV size, TSV spacing, and C4 spacing, we investigate the impact of coaxial TSVs and their novel usage in 3-D PDN.

4.2 Design Setup

4.2.1 3-D Stacked Architecture

We use a 3-D IC consisting of three dies: a quad-core chip-multiprocessor (PROC), a memory (MEM), and an accelerator engine (ACCL). Figure 4.2 shows the cross-section of our 3-D chip. Each die is assumed to have an area of $1 \, cm^2$. We consider the thermal/power profile of each die while considering their placement in the 3-D chip. Since PROC has the highest power consumption, we place it adjacent to the heat sink. We place ACCL farthest from the heat sink due to its lowest power consumption. MEM is placed at the center of the stack to allow shorter access paths from/to both PROC and ACCL. Each core of the CMP utilizes 10 W of maximum power, and is composed of five functional blocks: floating point unit (FPU), OOO (the rename, register file, result-bus, and window units), INT (integer arithmetic

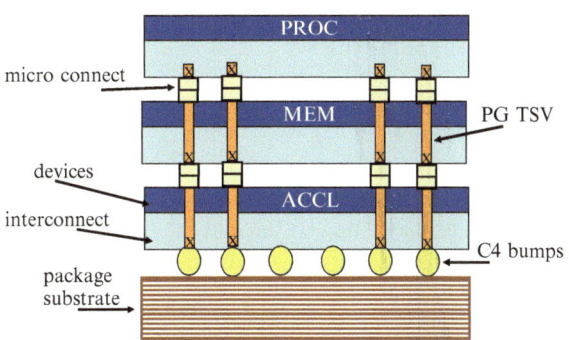

Fig. 4.2 Normal Stacked (3-D NOR) configuration

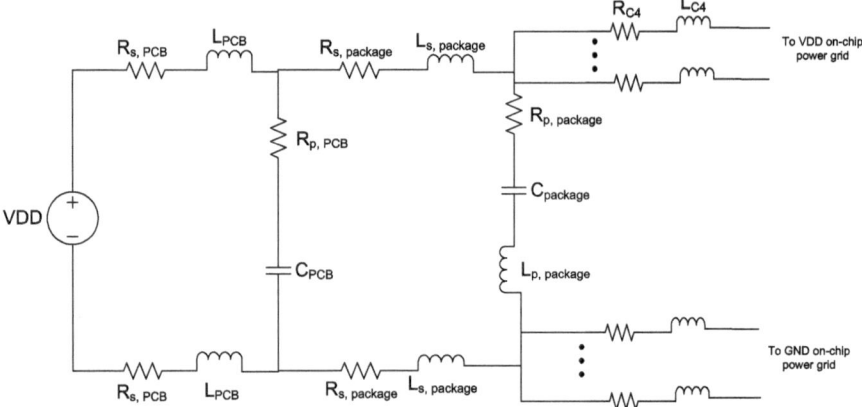

Fig. 4.3 Off-chip power delivery network

logic unit), Fetch (combines the instruction cache and branch predictor), and Data (represents the data cache and load-store queue). The MEM and ACCL modules utilize a maximum of 20 and 10 W, respectively. The maximum power consumed by the 3-D IC is 70 W. MEM is assumed to generate a current trace similar to that of the L2 Cache; ACCL is assumed to generate a current trace similar to that of the FPU block.

We use the regular square TSVs, as described in Sect. 2.1.1, for power delivery in the 3-D stack. To calculate the resistance and capacitance of individual TSVs and microconnects, we utilize the electrical characterization approach by Alam et al. [7], as described in Sect. 2.4. TSVs and microconnects provide connections for external I/Os and power delivery in the stacked chips. There are 16×16 VDD C4 connections per cm^2, and there is a similar number for the GND connections. Other vertical connections between the dies are used for inter-layer signal and thermal management. We refer to this configuration as a Normal Stacked (3-D NOR) Configuration to differentiate it from other 3-D configurations investigated later in the chapter.

4.2.2   Power Delivery Network (PDN)

The PDN model is illustrated in Figs. 4.3 and 4.4. The off-chip (motherboard and package), shown in Fig. 4.3, network is modeled as a resistive, inductive and capacitive network. The on-chip network, shown in Fig. 4.4, consists of a global-level, grid-like structure routed in top metal layers. We model the load imposed on the global grid as time varying current sources. The off-chip and on-chip networks are connected using series resistors and inductors representing the flip-chip package.

For the on-chip power grid, each grid element is modeled as a resistance and an inductance in series. In addition, current load points and microconnect or TSV

Fig. 4.4 A portion of the on-chip power grid for each die. Only select labeling is used

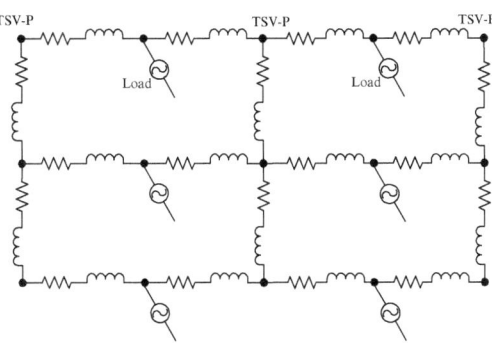

points (TSV-P) alternate throughout the grid as shown in Fig. 4.4. The TSV-Ps are connected to the C4 bumps either directly or through other stacked layers depending on the position of the on-chip PDN in the stack. The length of a grid element is such that we have a 16×16 element grid. We assume wide metal line widths such that the grid collectively occupies 50% of the total die area in the top two metal layers. We use the predictive technology model [3] to calculate the R & L for grid elements. A fast circuit solver, based on preconditioned Krylov subspace iterative methods [28], is used to solve the SPICE netlist for the modeled configuration. A decoupling capacitance of $33 \, \text{nF/cm}^2$ is assumed in our study, corresponding to device capacitance implementation with 1 nm gate oxide thickness (from the ITRS roadmap of 90 nm, 65 nm technology) occupying 20% of die area [46]. The decoupling capacitance is uniformly distributed along the grid elements in our 3-D IC.

4.2.3 Trace Selection

Predicting the power demand of the functional blocks is critical in evaluating the PDN performance. A simple strategy is to use the SPEC benchmarks to extract per cycle power demands of each functional block and analyze the PDN at each cycle. This strategy will give precise results but simulating the PDN for millions of cycles is time consuming. Another strategy is to extract the worst case power demands for each functional block and evaluate PDN performance for this worst case. While this strategy tremendously reduces the simulation time, the results are pessimistic as blocks typically do not operate at peak power at the same time, nor do they change current demand simultaneously. We use a trace compression strategy that reduces the simulation time without compromising the accuracy.

We use an architectural-level power model based on Wattch [21] to estimate the benchmark specific power dissipation in each functional block. Four SPEC benchmarks (apsi, bzip, equake, and mcf) are used to collect power traces for

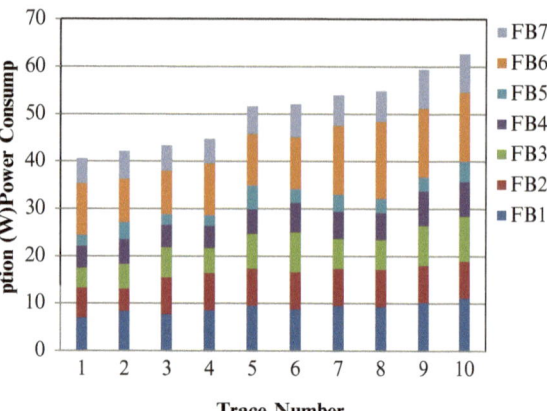

Fig. 4.5 Illustrative contents of a bin: the x-axis represents the trace number within a bin. The y-axis represents the total power dissipation in a particular clock cycle. The relative power distribution for functional blocks, FB1-FB7, is similar within each bin

millions cycles of each benchmark. We use the following strategies to extract the interesting traces for our IR and Ldi/dt analysis:

IR Analysis: For IR analysis, we divide all traces into bins such that each bin contains the traces with same distribution of normalized power dissipation over all the functional blocks. Figure 4.5 illustrates sample traces of one of the bins. While each trace in the bin contains a unique power demand for each functional block (FB1-FB7), the relative power dissipation of functional blocks, in each trace, is similar (within ±5% variations). Relative power consumptions of the functional blocks determine the capability of the PDN to share power for neighboring functional blocks. For each bin, we simulate only the trace with the largest total power consumption; for example, trace 10 will be simulated representing the bin shown in Fig. 4.5. We used this approach on 10 million traces of four benchmarks and we were able to reduce the number of traces to 1,428.

Ldi/dt Analysis: For Ldi/dt analysis, we use current traces that represent a variety of current patterns: step, pulse, and resonating. These patterns were derived based on the work of Meeta et al. [42], where four SPEC workloads (apsi, bzip, equake, and mcf) were run for 100 million instructions using Wattch [21], and 2,048 cycle snippets (8,192 total traces) representing the current patterns were then extracted. Such a power grid evaluation methodology replaces observing millions of instructions from a wide variety of benchmarks, thus significantly saving power grid simulation times.

4.3   Optimal TSV Size for 3-D PDN

We examine in this section how TSV size impacts 3-D power delivery. TSV size is the dimension of one side of the square TSV footprint. The TSV height is always equal to die thickness, which is 50 μm in all our 3-D setups. The maximum and

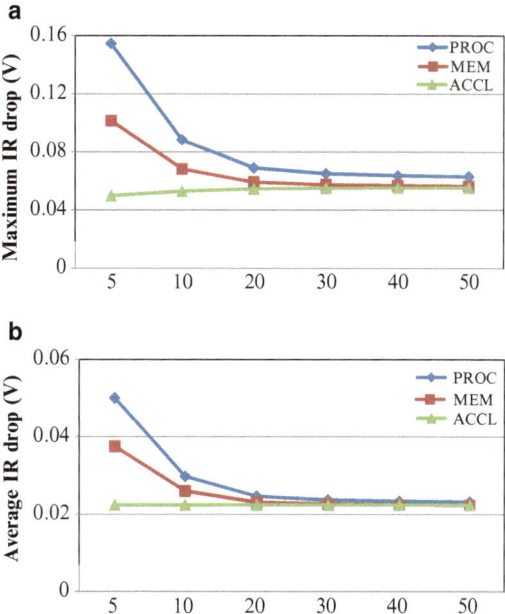

Fig. 4.6 Maximum and average IR drop for various TSV sizes. (**a**) Maximum IR drop, (**b**) Average IR drop

average IR drops in different dies, for TSV sizes ranging from 5 to 50 μm in the 3-D NOR configuration, are shown in Fig. 4.6. We can make the following observations:

- The ACCL, directly connected to C4 bumps, exhibits nearly constant maximum and average IR drops across the different TSV sizes. As TSV size is increased, there is a slight increase in both maximum and average IR drop. This represents the fact that the grid sharing capability is improved with the increased TSV size.
- More importantly, the IR drop saturates in PROC and MEM for TSV sizes of and greater than 20 μm. Such saturation suggests the lack of benefit of increasing the TSV size beyond a specific size. A TSV size of 25 μm is therefore used in the following analysis for 3-D PDN.

4.4 Best TSV Granularity

4.4.1 Baseline 3-D Configuration (3-D NOR)

This section presents IR and Ldi/dt analysis for the baseline 3-D configuration, described in Sect. 4.2.1. We use a TSV size of 25 μm. We run each of the compressed traces for IR and Ldi/dt as described in the Sect. 4.2.3 and observe the maximum,

Table 4.1 IR drop and Ldi/dt voltage droop for 3-D NOR configuration

	IR(V)			Ldi/dt(V)		
	PROC	MEM	ACCL	PROC	MEM	ACCL
apsi						
Maximum voltage drop	0.065	0.057	0.054	0.187	0.178	0.179
Average voltage drop	0.023	0.022	0.021	0.025	0.024	0.024
Standard deviation	0.008	0.007	0.007	0.019	0.019	0.019
bzip						
Maximum voltage drop	0.066	0.058	0.055	0.325	0.313	0.314
Average voltage drop	0.023	0.022	0.022	0.061	0.060	0.061
Standard deviation	0.008	0.007	0.007	0.041	0.041	0.041
equake						
Maximum voltage drop	0.065	0.056	0.053	0.334	0.328	0.310
Average voltage drop	0.025	0.024	0.023	0.035	0.034	0.033
Standard deviation	0.008	0.007	0.007	0.036	0.036	0.036
mcf						
Maximum voltage drop	0.065	0.057	0.054	0.341	0.347	0.346
Average voltage drop	0.025	0.024	0.023	0.040	0.040	0.039
Standard deviation	0.007	0.007	0.006	0.029	0.029	0.028
Across all 4 benchmarks						
Maximum voltage drop	0.066	0.058	0.055	0.297	0.292	0.287
Average voltage drop	0.024	0.023	0.022	0.040	0.040	0.039
Standard deviation	0.008	0.007	0.007	0.031	0.031	0.031

average, and standard deviation. The results for this analysis are presented in Table 4.1. These results form our baseline case, and all other analyses in this chapter are presented in reference to these values. Examining the results presented in Table 4.1, it is clear that the 3-D architecture is not an ideal one because the maximum Ldi/dt voltage droop are considerably higher than their averages. We assume that local adjustments, in the form of adding local decoupling capacitors, can be done for 3-D power delivery networks. We therefore keep the architectures and power delivery network parameters the same in all the studies presented in this chapter.

4.4.2   Effects of TSV Granularity (Spacing)

The PDN in 3-D NOR configuration is assumed to have the same TSV spacing as that of the C4 connections. To design a 3-D stacked configuration that enables increasing the granularity of TSVs for power distribution in any of the dies in the 3-D stack, we introduce an interposer die [83] between the C4 connections and the bottom die as illustrated in Fig. 4.7. The interposer acts as a redistribution layer that is connected to C4 bumps on one side and bonded microconnects (higher granularity) on the other, thus distributing power to the top die via the TSVs. We can

Fig. 4.7 Stacked Interposer
(3-D SI) configuration

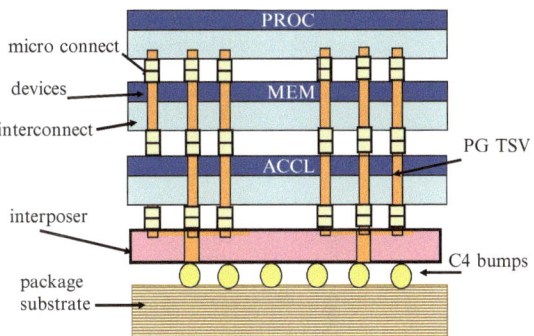

therefore decrease the TSV spacing in the PDN to as low as the minimum allowed microconnect pitch while the C4 pitch can remain unchanged. We refer to this setup as the Stacked Interposer (3-D SI) configuration. We keep the granularity of C4 bumps (16×16 connections per cm^2) the same as that in 3-D NOR configuration. The off-chip power delivery network also remains unchanged.

To assess the impact of the TSV spacing on power delivery, we vary the TSV granularity from 16×16 in our 3-D NOR configuration to granularities: 32×32, 48×48, and 64×64. At the highest granularity of 64×64, the TSV spacing is well above the minimum TSV pitch limit of $0.4 \mu m$ in wafer-to-wafer and of $5 \mu m$ in die-to-wafer or die-to-die 3-D bonding technologies [107]. The silicon area consumed by TSVs in the 3-D PDN for the 32×32 and 64×64 granularities are 5% and 20%, respectively, for a TSV size of $25 \mu m$. For each increased granularity, the physical dimensions of each grid element are adjusted, and R and L values are recalculated. The decoupling capacitance is uniformly redistributed throughout the on-chip grid on each die and its total value remains same.

Table 4.2 reports the results from static IR and transient Ldi/dt voltage analysis with various TSV spacing in the 3-D SI configuration. All values are normalized to the 3-D NOR architecture. We report the results across all benchmarks. We make the following observations:

• Despite the expectation that increasing TSV granularity in the 3-D PDN would improve the overall performance of power delivery, we notice only marginal improvements in all the metrics for IR drop. The maximum IR drop in the PROC die is improved only 8% by increasing the TSV granularity from 32×32 to 64×64 whereas the TSV silicon area penalty rises from 5 to 20%. Similar observations are made for the transient voltage droop where the improvements in the maximum and average voltage droop figures are 9% and 3%, respectively.

• The marginal improvement suggests that an on-chip grid and TSV granularity of 32×32 reaches a near optimum solution for power grid performance, particularly for IR drops. This observation leads us to consider improving the off-chip network by examining the granularity of C4 bumps, which we explore next.

Table 4.2 IR drop and Ldi/dt voltage droop analysis for different TSV granularities in 3-D SI architecture. The results are normalized to the 3-D NOR values

	IR(V)			Ldi/dt(V)		
	PROC	MEM	ACCL	PROC	MEM	ACCL
32×32						
Maximum voltage drop	0.94	1.04	1.09	0.89	0.88	0.88
Average voltage drop	0.96	1.00	1.01	1.00	1.01	1.03
Standard deviation	0.93	0.98	1.03	1.10	1.11	1.12
48×48						
Maximum voltage drop	0.94	1.06	1.12	0.99	0.98	0.98
Average voltage drop	0.95	1.00	1.02	0.96	0.98	1.00
Standard deviation	0.92	0.99	1.05	0.98	0.99	1.00
64×64						
Maximum voltage drop	0.86	0.97	1.03	0.98	0.98	0.98
Average voltage drop	0.94	0.98	1.00	0.97	0.98	1.01
Standard deviation	0.89	0.96	1.02	0.87	0.88	0.89

Table 4.3 IR drop and Ldi/dt voltage droop analysis for 3-D NOR with both C4 and TSV granularities of 32×32. The results are normalized to the 3-D NOR values

	IR(V)			Ldi/dt(V)		
	PROC	MEM	ACCL	PROC	MEM	ACCL
Maximum voltage drop	0.55	0.59	0.60	0.90	0.89	0.89
Average voltage drop	0.61	0.63	0.64	0.86	0.87	0.89
Standard deviation	0.59	0.62	0.64	1.04	1.05	1.06

4.4.3   Effects of C4 Granularity (Spacing)

We now assume that the 3-D NOR configuration with both C4 and TSV having equal granularity of 32×32 for both VDD and GND supply networks. This is an increase over the 16×16 C4 granularity used earlier. We perform IR and Ldi/dt analysis, and summarize the results in Table 4.3. We make the following observations:

- Increased C4 granularity results in significant improvement in IR voltage drop. This 4× increase in the number of TSV and C4 results in improved 3-D PDN performance.
- Although increasing C4 granularity significantly improves IR drops, the improvement is limited in terms of Ldi/dt voltage droop. This is due to the off-chip PDN components (package and PCB) having a more dominant impact on Ldi/dt voltage droop.

4.5 Effect of Dedicated Power Delivery

The experiments in the previous section assume that TSVs in the 3-D PDN are shared among all dies. In this section, we study the effect of adding partially dedicated power delivery to each die through a few TSVs connected to only select dies. We define a new 3-D configuration, the tapered stacked (3-D TAP) Configuration, shown in Fig. 4.8.

In the 3-D TAP configuration, dies are progressively sized larger to be able to connect few dedicated vertical connections, called the boundary vias, to the PDN in the extended boundary portion of a die. As illustrated in Fig. 4.8, the boundary vias do not pass through any of the active silicon area and can be formed using advanced package-level routing vias similar to those in redistributed chip packaging [4]. The size of each die is modified such that the tapering ratio is constant between the dies and a total silicon area of $3\,cm^2$ is achieved for the 3-D chip. Due to die resizing, we modify the module placements: the top die now has PROC and some part of ACCL, the middle die has MEM, and the bottom die has ACCL. Parameters for C4 pitch, on-chip power grid R & L, and off-chip network are kept the same as in the 3-D NOR configuration. Due to an increase in footprint, there is an increase in the number of C4 and TSVs of 18×18 compared to 16×16 in 3-D NOR configuration.

The results for IR drop and Ldi/dt voltage droop analysis in 3-D TAP configuration are presented in Table 4.4. The results show that partly dedicated power delivery in 3-D TAP configuration does not have the same extent of improvement as increasing C4 granularity (comparing the results to those in Table 4.3). However, both average IR drop and Ldi/dt voltage droop in the 3-D TAP configuration are

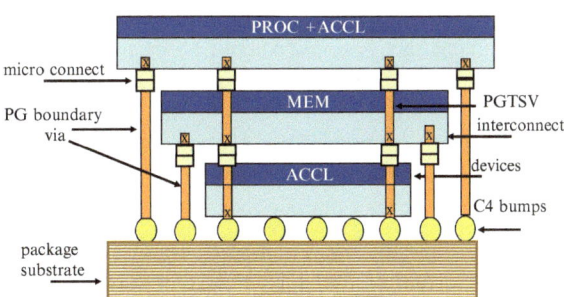

Fig. 4.8 Tapered Stacked (3-D TAP) configuration

Table 4.4 IR drop and Ldi/dt voltage droop analysis for 3-D TAP configuration. The results are normalized to the 3-D NOR values

	IR(V)			Ldi/dt(V)		
	PROC	MEM	ACCL	PROC	MEM	ACCL
Maximum voltage drop	0.84	0.81	0.80	0.98	0.98	0.99
Average voltage drop	0.74	0.75	0.74	0.94	0.96	0.98
Standard deviation	0.92	0.95	0.97	0.97	0.97	0.98

improved compared to the 3-D SI (Table 4.2, 32×32 TSV granularity case). This improvement does not have any silicon area penalty as in the 3-D SI because TSV granularity is the same as in the 3-D NOR configuration.

Although the concept of dedicated or partly dedicated power delivery in 3-D IC, as in 3-D TAP, is interesting and effectively improves performance of 3-D PDNs, there may be additional risk and cost considerations associated with tapered die sizing and non-standard boundary via packaging process. The tapered die sizes would only permit die-to-die and die-to-wafer bonding techniques excluding the wafer-to-wafer option which requires the same die and wafer sizes. Process considerations aside, the 3-D TAP configuration illustrates a method for isolating some of the most active parts of dies by using dedicated delivery in that area. The proposed method also yields improvement in 3-D PDN vis-a-vis other 3-D PDN configurations.

4.6   Power Delivery Using Coaxial TSV

Coaxial TSVs, as described in Sect. 2.1.4, are proposed to eliminate substrate noise by grounding the outer metal layer while the inner metal layer is used for signal transmission. We investigate in this section using coaxial TSVs for power delivery for reducing blockages, increasing decap, and overlaying power/signal routing.

4.6.1   Reducing Blockages

When power TSVs extend through a die, routing blockages are created in the x-, y-, and z-dimensions. A single coaxial TSV can deliver VDD and GND simultaneously, reducing the overall number of power supply TSVs. We consider an architectural setup similar to the 3-D NOR configuration described in Sect. 4.2.1 for IR drop and Ldi/dt voltage droop analysis. We keep the cross-section for each of the inner and outer metal layers the same as the square-TSV area used in previous analyses. We assume an inner liner thickness of 20 nm, resulting in an overall size of each coaxial TSV of width 35.4 μm. TSV granularity of 16×16 for each VDD and GND supply with square TSV is now translated to coaxial TSV granularity of 16×16 used for delivering both VDD and GND. While our analysis suggests that this setup does not improve the IR drop or Ldi/dt voltage droop, merging two square-TSVs into single coaxial TSV results in a fewer number of routing blockages.

4.6.2   Increasing Decap

A coaxial TSV of width 35.4 μm with an inner liner thickness of 20 nm has a capacitance of 9.96 pF. This TSV capacitance is approximately 250× the square-TSV capacitance and acts as additional decap in the 3-D PDN. While the 20 nm inner liner thickness is selected for illustrative purpose, further process development work is ongoing for controlling the inner liner thickness to effectively implement capacitors using coaxial TSVs [8]. In this experiment, we analyze how to exploit coaxial TSVs to maximize on-chip decoupling capacitance in 3-D PDN.

We replace each square VDD or GND TSV with a coaxial TSV with a thin outer metal layer (example thickness of 0.2 μm) connected to the opposite power rail. Thus, the granularities of VDD and GND TSVs do not change while each TSV gets additional decoupling capacitance due to the co-axial implementation. Choosing an inner liner thickness of 20 nm, the coaxial TSV will have approximately 16% of the original decoupling capacitance. While the 16% increase in decoupling capacitance does not translate to significant improvement in Ldi/dt voltage droop as seen in our analysis, the coaxial decoupling capacitance implementation can be viewed as an opportunity to free silicon area by reducing the silicon area required for implementing device decoupling capacitance. In this example, device decoupling capacitor area will reduce by 16% when using a coaxial TSV implementation.

To implement a large amount of decoupling capacitance in coaxial TSVs, researchers proposed an alternative implementation with more than one layer of inner liner and inner metal such that multiple inner liner layers collectively form a large capacitor [97]. Using such multilayer coaxial TSV, if we increase decoupling capacitance by 90%, then the peak Ldi/dt noise improvement from our analysis is approximately 15%. Other researchers also proposed adding large amounts of decoupling capacitance, such as 80% more decoupling capacitor implemented as additional decap die stacked in 3-D chips, which provided 22–36% peak Ldi/dt noise reduction [46]. However, additional decap die would block cooling paths for other dies. Therefore, a coaxial TSV can be considered as a new alternative for decoupling capacitor implementation in 3-D PDN with an opportunity to improve Ldi/dt voltage droop or save silicon area by replacing device capacitors.

4.6.3 Overlaying Signal and Power Routing

When the resistance of a particular power TSV, for example one in non-hotspot area, is not highly critical, a coaxial TSV of the same footprint can be used to route an additional (non-VDD and non-GND) signal. Signal TSVs are expected to be smaller in size for reduced capacitive load making them ideal to overlay with large size power TSVs that use a coaxial TSV implementation. In this experiment, we study the effect of power/signal overlay in a 3-D PDN using coaxial power TSVs. The coaxial power TSV of width 25 μm with the inner metal of width 5 μm is dedicated for signal transmission and the outer metal is used to deliver power. We incorporate this new coaxial TSV scheme into the 3-D NOR configuration and perform IR drop and Ldi/dt analysis. Note that coaxial TSV footprint is the same as that of the core TSV footprint in the 3-D NOR configuration. Each one of the VDD and

Table 4.5 Summary of co-axial TSV analysis for 3-D PDN

	No. of blockages	Size of each blockage	Additional signal routes due to overlay	Benefits
TSV placement	256	$1.253\,E-9m^2$	0	Reduced number of blockages
Decap insertion	512	$0.841\,E-9m^2$	0	Additional decoupling capacitance
Signal overlay	512	$0.600\,E-9m^2$	512	Additional signal routing

GND coaxial TSVs is used to transmit a signal through the inner metal layer which would represent the extreme case of signal overlay for analyzing worst case impact on PDN performance.

We did not observe any change in IR drop or Ldi/dt voltage droop of the 3-D PDN with overlay coaxial TSVs. This is due to the fact that we sacrificed less than 4% of the TSV area for signal overlay. This observation is also supported by the IR saturation trend in Fig. 4.6 where we notice that a TSV size of 25 μm is already near saturation. Hence, reducing effective power TSV area by a small percentage did not significantly degrade the performance of the 3-D PDN.

Table 4.5 summarizes the above presented three studies to evaluate the potential benefit of using coaxial TSV for 3-D PDN. The first column shows the technique used to integrate coaxial TSV into the 3-D PDN. The next three columns quantify three parameters (number of blockages, size of each blockage, and the number of additional signal routes) that we use to compare the implementations. The last column summarizes the main benefit obtained from each integration technique. Clearly, coaxial TSVs present an exciting opportunity to reduce the number of blockages, integrate extra decoupling capacitance, and provide additional signal routes. All these benefits have no extra area or performance penalty.

4.7 Best Practices for 3-D PDN Design and Optimization

Based on our findings, we present a set of guidelines for designing and optimizing power delivery networks in future 3-D designs.

- A critical observation in our work is the saturation trend of IR drop in 3-D PDNs with increased TSV size. This suggests the need for finding the optimal TSV size and grid placement given the on-chip grids in 3-D stacked layers such that the least amount of silicon penalty is incurred.
- Increasing TSV granularity or equivalently decreasing TSV spacing in 3-D PDN improves the standard deviation in IR drop and Ldi/dt voltage droop most, with

marginal improvements in maximum and average values. Therefore, physical design for 3-D PDN must consider the impact and choose TSV granularity for minimum silicon area penalty.

- Despite selecting the optimal TSV size and TSV spacing, 3-D PDN performs worse in both IR drop and Ldi/dt voltage droop if the package connection, such as C4, pitch or granularity is maintained the same. Our study shows that improving off-chip component of the 3-D PDN, for example through reducing the C4 pitch for a higher number of C4s, has the highest relative impact on power grid metrics.
- A combination of shared and dedicated TSV power delivery can be used, as illustrated in 3-D TAP configuration, to achieve improvements in both IR drop and Ldi/dt voltage droop. Further investigation is recommended with dedicated power grid approaches for physical design, such as floorplanning and placement of dedicated TSVs, for optimization of such 3-D PDNs.
- Coaxial TSVs provide unique opportunities for overlaying power supply routes, providing additional decoupling capacitance, and overlaying power/signal routes.

The above guidelines apply to power delivery networks that use C4 and TSVs for the proposed 3-D power delivery networks. For other 3-D PDN designs within different packaging, separate analysis will be required to determine an appropriate set of guidelines.

4.8 Summary

Power delivery will be a major physical design concern in 3-D ICs due to higher power density and package asymmetries. We performed in this chapter a detailed analysis to study the IR drop and Ldi/dt voltage droop in the context of various design parameters in 3-D PDNs. As TSVs occupy valuable die real estate, we analyzed the impact of TSV size and spacing to analyze the trade offs between TSV area and PDN performance. Our results show that PDN performance saturates for a TSV size of $25\,\mu m$ and a TSV granularity of 32×32. We studied off-chip PDN (increased C4 granularity) and a combination of shared and dedicated TSVs to improve PDN performance. A $4\times$ increase in C-4 granularity reduces the average IR drop by 39% and average di/dt droop by 14%. Using dedicated TSVs, reduces average IR drop and di/dt droop by 26% and 6%, respectively. In addition, we show that coaxial TSVs improve PDN performance by providing additional decoupling capacitance, and that coaxial TSVs reduce interconnect and device blockage by overlaying power/signal routing within a single coaxial TSV.

Chapter 5
Early Estimation of TSV Area for Power Delivery in 3-D ICs

To harness the full potential of 3-D integrated circuits, analysis tools for early design space exploration are needed. Such tools, targeting multiple design facets and cost trade-off analysis, would allow designers to arrive at major decisions regarding architecture and implementations fabrics. TSVs occupy valuable silicon real estate and neither devices nor interconnects can be formed in the area occupied by a TSV. An early estimation of total TSV area allows effective budgeting of device and interconnect resources.

We provide in this chapter a set of algorithms to estimate the minimal substrate area required for TSV-based power delivery. These algorithms can be applied early in the design stages when only functional block-level behaviors and a floorplan are available. Our proposed work is in contrast with recent TSV optimization techniques, which are applied later in the design cycle [62, 120], to target optimal signal TSV placement. Planning for signal TSVs clearly requires detailed circuit layout. Our optimization framework can be extended for investigating using TSVs for heat removal if realistic thermal modeling is appropriately utilized. Studies supporting the use of TSVs for heat removal are still in their early stages and do not consider realistic manufacturing constraints.

The rest of the chapter is organized as follows: Related work is presented in Sect. 5.1. The problem of power TSV area minimization is formulated in Sect. 5.2. Our proposed algorithms are presented in Sect. 5.3. We discuss our results in Sect. 5.4. We summarize our work in Sect. 5.5.

5.1 Related Work

To optimize TSV area, Lee et al. investigated a co-optimization methodology for signal, power, and thermal TSVs based on design of experiments and response surface method, and they showed that careful tuning of response surface models can lead to reliable optimization results [62]. Yu et al. use I/O compression and

N. Khan and S. Hassoun, *Designing TSVs for 3D Integrated Circuits*, SpringerBriefs in Electrical and Computer Engineering, DOI 10.1007/978-1-4614-5508-0_5, © The Authors 2013

structured and parametrized model order reduction to efficiently ensure dynamic power/thermal integrity [120]. Current TSV co-optimization techniques thus provide insights but have neglected key TSV manufacturing constraints. While functional data early in the design cycle can be used to estimate thermal profiles, studies addressing the manufacturing concerns are needed to further explore the benefits of thermal TSVs. Each TSV must have a *liner*, an insulating material filled around the TSV, to provide isolation as well as stop metal diffusion into substrate. This insulating material helps to reduce TSV-induced substrate noise but deteriorates TSV performance to extract heat from substrate. A model of thermal TSV without liner is, thus, not accurate and none of the previous models of thermal TSV include liner. Mechanical stress associated with TSV calls for a keep-away area where no devices can be fabricated. Therefore TSVs cannot be directly connected to a hotspot or any floorplan tile as assumed in [120]. Power TSVs connect to only the top metal layers. The efficiency of heat extraction from the surrounding volume is unclear.

While this chapter investigates power TSV optimization, independently of thermal TSVs, the iterative framework can be extended to address co-optimization if realistic manufacturing assumptions are available.

5.2 Problem: Power TSV Area

Because the proposed TSV estimate occurs early in the design cycle, detailed device-level floorplans are unavailable. At such early stages, a functional model of each die is available. Thus, a set of workloads (e.g., for the target design) can be executed using an architectural simulator and the power traces of each functional block are captured. Within each functional block, we assume uniform power consumption.

We assume a 3-D IC consisting of K number of dies in a flip chip package. Each die has its own on-chip power grid, each with M grid nodes. The bottom die is connected to an off-chip PDN via C4 bumps and the rest of the dies are interconnected using TSVs. Because our technique targets early design exploration, we assume uniform TSV sizing and that TSV insertion points have already been identified, each with an index, $1 \leq i \leq M$. Each TSV grid location is referred to as a TSV node t_i. Figure 5.1 illustrates a 2×2 portion of an on-chip power grid.

The size of a power TSV depends on several factors including fabrication process, power delivery requirements [52], stress minimization [109], heat removal requirements [120], and layout constraints. 3-D power grid design studies recommend larger power TSVs (with lower resistance) to reduce voltage drops and to meet current density requirements [46,52]. However, larger TSV sizes directly impact the keep-away area, an area around each TSV where no devices or interconnects can be fabricated. Bart et al. suggest that the keep-away area increases with the increase in TSV area [109]. A simple calculation based on their findings shows that for $1.7\times$ increase in the TSV area (increase in diameter from 6 to 8 μm), a $2.4\times$ increase in keep-away area is required. So, instead of using few larger TSVs, we use multiple

Fig. 5.1 An illustrative 2×2 power grid

Fig. 5.2 Side and top view of
select power delivery nodes
in a stack of three dies,
illustrating a device node's
neighborhood

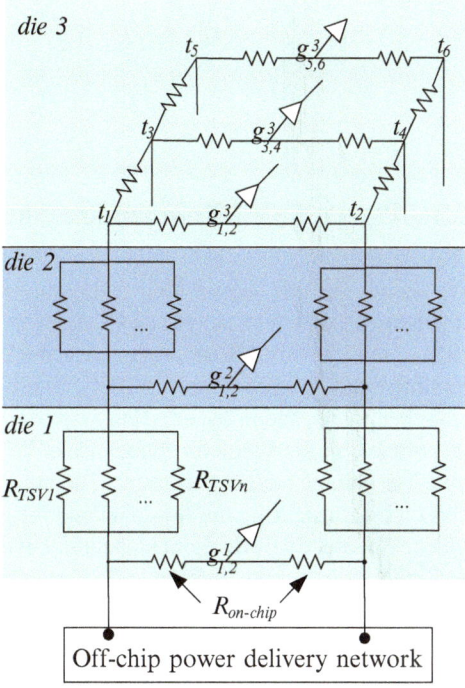

smaller TSV arrangements connected in parallel between two TSV grid points. For
this paper, we assume a TSV diameter of 5 μm. Each location t_i is assigned a number
of TSVs, n_i.

We assume that power TSVs supply neighboring devices as shown in Fig. 5.2.
Each device node, $g_{i,j}^k$ connects the two closest TSV grid locations, t_i and t_j, and
the device node is located on die k. TSVs at locations t_i and t_j are thus directly
connected to $g_{i,j}^k$. We define the neighborhood of a device node as the set of 6 TSV
locations closest to $g_{i,j}^k$. This is illustrated simply in die 3 of Fig. 5.2.

We denote the voltage at any node $g_{i,j}^k$ as $v_{i,j}^k$. Voltage violation is defined as the difference between $(V_{ref} - V_b)$ and $v_{i,j}^k$, where V_{ref} is the reference (input) voltage and V_b is the maximum allowable noise voltage. Voltage slack, on the other hand, is defined as the difference between $v_{i,j}^k$ and $(V_{ref} - V_b)$.

The total power TSV area can be calculated as $S \times \sum n_i$, where S is the TSV size and n_i is the number of TSVs at each node t_i. Our objective is to minimize the total number of power TSVs under the constraint that power integrity is maintained.

The power TSV minimization problem can be formulated as follows:

$$\min \sum_{i=1}^{M} n_i$$

subject to:

$$\forall_{i,j,k} : \left| V_{ref} - v_{i,j}^k \right| \leq V_b$$

$$\forall_i : n_i \geq 0$$

5.3 Power TSV Minimization Algorithms

We propose in this section four techniques (REDUCE MAXIMUM SLACK (RMS), REDUCE SOMEWHAT ARBITRARY SLACK (RSAS), REDUCE SLACK LOCALLY (RSL), and IMPROVE WORST VIOLATION (IWV)) within an iterative framework to minimize the number of power TSVs. These techniques differ in their starting and stopping points, and in the decision made during each iterative step. In the first three techniques, we start with an abundance of TSVs allowing the circuit to comfortably meet the voltage budget constraint. The initial number of TSVs at each grid node is chosen by considering the floorplan and the power requirements of each functional block. We selectively decrement the number of TSVs and continue till all efforts fail to further decrease the number of TSVs while maintaining power integrity.

In contrast to the first three techniques, IMPROVE WORST VIOLATION (IWV) starts with a scarcity of TSVs and judiciously increases the number of power TSVs during each iterative step. A small number of TSVs is initially assigned to each node, selected based on the designers' experiences, or simply with one TSV. During the iterative process, the number of TSVs at the node with the largest voltage violation is incremented. The process repeats until all nodes meet the required noise budget.

5.3.1 Reduce Maximum Slack (RMS)

We explore in this technique the effectiveness of decreasing the number of TSVs for a node with the largest slack. An initial circuit is generated assuming uniform n_i for each grid node. This initial estimate considers the power demand of each functional

Input: An initial design with an abundance of TSVs
Output: The minimum n_i for each grid node such that each node meets the voltage budget

1 mark all device nodes $g_{i,j}^k$ "not done"
2 initialize *noOfNodesDone* to zero
3 run SPICE
4 **while** *noOfNodesDone* $< M$ **do**
5 | let g_s denote the device node that is "not done" and has largest voltage
6 | pick node t_x which is g_s's direct TSV neighbor that has the largest slack
7 | decrement n_x by 1
8 | run SPICE
9 | **if** *circuit fails* **then**
10 | | increment n_x by 1
11 | | pick the second neighboring TSV node t_y
12 | | decrement n_y by 1
13 | | run SPICE
14 | | **if** *circuit fails* **then**
15 | | | increment n_y by 1
16 | | | mark g_s "done"
17 | | | increment *noOfNodesDone* by 1;
18 | | **end**
19 | **end**
20 **end**

Algorithm 1: REDUCE MAXIMUM SLACK (RMS)

block. We assign a small number of TSVs (one, if no other information is available), at the TSV grid nodes within the functional blocks with the smallest power demand. We then assign a relative n_i to rest of the grid nodes in the other functional blocks. We run SPICE assuming power requirements (as explained in our trace selection methodology in Sect. 4.2.3). If any node $g_{i,j}^k$ fails the voltage budget constraint, we increment the minimum n_i for all nodes. We reevaluate the circuit. We continue this process until all $g_{i,j}^k$ meet the voltage budget constraint.

Pseudo code for the RMS iterative algorithm is presented in Algorithm 1. After initialization, and at the beginning of each iteration of the while loop, a device node $g_{i,j}^k$ with the maximum voltage is identified as g_s (line 5). Then, the direct TSV neighbor node with the largest voltage slack, t_x, is identified (voltage measurements taken from the furthest away from C4 bumps, typically the worst case voltages). The number of TSVs at t_x, n_x, is decremented by one (line 7), and SPICE is used to evaluate the results (line 8). If the circuit fails, then the algorithm attempts to reduce the number of TSVs at the other direct neighbor, t_y, while restoring the TSV count at t_x. If both attempts fail, then the identified node g_s is marked as "done" (line 16), indicating that the device node cannot withstand the downsizing of its direct TSV neighboring nodes. The algorithm stops when all the device nodes are marked as "done", and no further TSV decrease is possible.

```
 1  while noOfNodesDone < M do
 2  │    pick a random device node g_s that is "not done"
 3  │    let S be a set of TSV nodes in g_s's neighborhood
 4  │    while S is not empty do
 5  │    │    find t_s s.t. t_s ε S and t_s has maximum slack
 6  │    │    decrement n_s by 1
 7  │    │    run SPICE
 8  │    │    if circuit fails then
 9  │    │    │    increment n_s by 1
10  │    │    │    remove t_s from S
11  │    │    end
12  │    end
13  │    mark g_s "done"
14  │    increment noOfNodesDone by 1;
15  end
```

Algorithm 2: REDUCE SLACK LOCALLY (RSL)

5.3.2 Reduce Somewhat Arbitrary Slack (RSAS)

This technique is similar to RMS, but differs in which of g_s's direct neighbors will be selected for TSV decrementing first. In particular, the algorithm differs in line 6, where the code is changed to select a direct neighbor at random. The rational is to avoid a purely greedy technique as was the case in RMS.

5.3.3 Reduce Slack Locally (RSL)

To further move away from a greedy technique, we explore in this technique the impact of minimizing the TSV count at nodes other than direct neighbors, and the impact of not immediately selecting the nodes with the maximum slack. The algorithm initialization is similar to the one shown in Algorithm 1, but the while loop is different and is presented in Algorithm 2. A node whose local neighborhood was not thoroughly explored is randomly selected. Within the neighborhood, TSV nodes with the maximum voltage slack are chosen for decrementing successively until none of the neighborhood TSV nodes can be further decremented. Once a neighborhood is explored, a device node is marked as "done". The iterations stop when all the device nodes are marked as "done".

5.3.4 Improve Worst Violation (IWV)

While the first three techniques attempted to lower an abundance of TSVs at each TSV node, IWV starts with few TSVs and increases the number of power TSVs during each iterative step. The pseudo code is shown in Algorithm 3. The initial

Input: An initial circuit with a low number of TSVs
Output: The minimum n_i for each grid node such that each node meets the voltage budget
1 run SPICE
2 Find the minimum voltage, v_s, in the circuit
3 **while** $v_s < (V_{ref} - V_b)$ **do**
4   pick a device node g_s s.t. its voltage is equal to v_s
5   let S be a set of TSV nodes in g_s's neighborhood
6   find t_s s.t. $t_s \varepsilon$ S and t_s has the largest noise violation
7   increment n_s by 1
8   run SPICE;
9 **end**

Algorithm 3: IMPROVE WORST VIOLATION (IWV)

circuit has a small number of TSVs at each grid node, selected based on experience, or simply with one TSV. During the iterative process, the number of TSVs at the node with the largest voltage violation is incremented. The process repeats until all nodes meet the required noise margins. This algorithm is greedy as well, as it always tries to improve the voltage of the most offending device node by incrementing the TSV count at the TSV node with the lowest voltage.

5.4   Results and Discussion

We ran the four algorithms to estimate the number of TSVs for a block-level model of a 3-D IC that is similar to the one used in Chap. 4, where we derive values of on-chip and off-chip power delivery components from published works and technical documentation. We utilize the electrical characterization approach by Alam et al. [7] to calculate the resistance of individual TSVs.

Figure 5.3 shows the estimated number of TSVs for the selected traces of one of the benchmarks, bzip. The results obtained using IWV are always better than the other three algorithms. Each increase in the number of TSVs in IWV brings the device nodes closer to meeting the voltage requirement. The process is incremental. For the other three techniques, the minimization is a two-step process: a device node is selected and then a TSV node is chosen to decrease the number of TSVs. This strategy leads to a state where n_i for a node is decreased to an extent such that any decrement anywhere else in the circuit results in circuit failure. The greedy nature of the algorithms does not allow any significant backtracking to explore other options. Among the three reduction techniques, RMS and RSAS result in similar TSV numbers. The results of RSL however were inconsistent. The TSV estimates were sometimes much worse than the ones obtained by RMS and RSAS, and sometimes better.

The estimated minimum required number of TSVs at each TSV node t_i is the maximum value of n_i across all benchmarks as the n_i choice must satisfy the demands of the worst-case benchmark. The total number of TSVs required

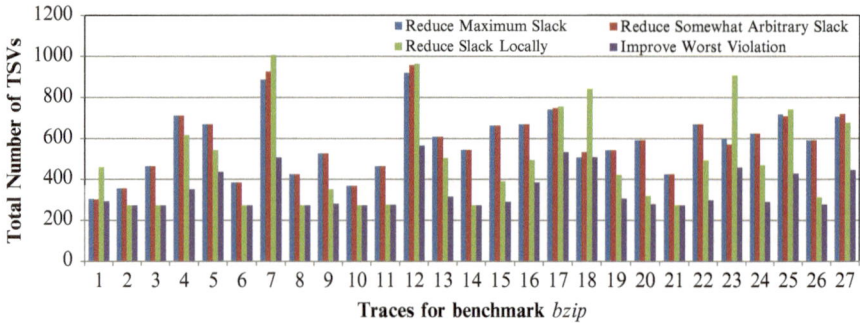

Fig. 5.3 The total number of TSVs required for each select trace for the bzip benchmark using the four minimization techniques

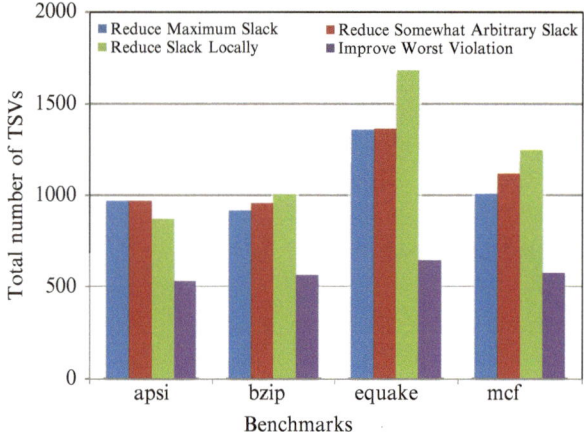

Fig. 5.4 Comparison of the four proposed techniques

by each benchmark is shown in Fig. 5.4. Clearly, IWV provides the best and smallest results for all four benchmarks. RMS and RSAS provide similar estimates. The performance of RSL is less consistent and dependent on the benchmark. For comparison, we ran these algorithms for a pessimistic power dissipation scenario, assuming worst case power dissipation of each functional block across all benchmarks. The total TSV count was $\approx 2.7\times$ the minimal TSV count found using select traces.

The run time of the iterative algorithms is dependent on the network details and number of insertion points, with SPICE as the bottleneck. The iterative algorithm can be run for each of the traces independently on a different machine. The run time thus becomes feasible. For a sample circuit with functional blocks of three dies (processor, memory, and accelerator chips) and 256 TSV grid points, the algorithm requires about one hour on a cluster/farm with 40 processing nodes, common in circuit design houses.

5.5   Summary

We have developed and evaluated several estimation techniques to evaluate power TSV area requirements for 3-D power delivery when only a functional model and a floorplan are available. We have shown that an iterative framework is feasible. Within this framework, we have shown that a greedy algorithm that gradually increments the number of TSVs at each grid node provides a minimal solution when compared to techniques that start with an abundance of TSVs and attempt to lower the TSV count. For a 3-D IC containing three dies, the incremental approach provides at least $0.6\times$ smaller TSVs than any of the other approaches. The run time of this iterative procedure can scale to one hour when dispatched on a cluster of 40 processing nodes.

Chapter 6
Carbon Nanotubes for Advancing TSV Technology

.

Resistivity, current density, and electromigration create concerns regarding the viability of copper (Cu) for interconnect in future CMOS technologies [13, 20, 70]. Carbon Nanotubes (CNTs) have emerged as a promising technology to meet these challenges [12, 78, 99]. We investigate in this chapter the feasibility of using CNTs for power delivery in a 3-D IC. We evaluate CNTs as TSVs and as on-chip power grid. We demonstrate that using CNTs in the power delivery network (PDN) of a 3-D IC results in tremendous reduction in the area occupied by TSVs for power delivery. Power can be delivered at coarser granularity while using CNT, thus requiring less number of TSVs and resulting in smaller chip area dedicated to TSVs.

The rest of the chapter is organized as follows: We present an overview of CNTs and some relevant research in Sect. 6.1. We describe our design setup and different design configurations in Sect. 6.2. We evaluate the use of CNTs to reduce the substrate area dedicated to TSVs for power delivery in a 3-D IC in Sect. 6.3. We present static IR analysis for different design configurations in Sect. 6.4. We evaluate CNTs for dynamic Ldi/dt performance in Sect. 6.5. We summarize our work in Sect. 6.6.

6.1 Carbon Nanotubes (CNTs)

Carbon nanotubes are hollow cylinders composed of one (single-walled CNT) or more (multi-walled CNT) concentric layers of carbon atoms in a honeycomb lattice arrangement. The particular electronic structure of the lattice determines if CNT is conducting (metallic), or semi-conducting. CNTs can be grown to a few centimeters in length [122]. The diameter of a single-walled CNT (SWCNT) can range from 0.5 nanometer to a few nanometers while the diameter of a multi-walled CNT (MWCNT) can range from a few to more than one hundred nanometers [78].

Superior-strength carbon–carbon bonds and near-perfect side-wall structures make CNTs highly reliable and much more resistant to electromigration than

N. Khan and S. Hassoun, *Designing TSVs for 3D Integrated Circuits*, SpringerBriefs in Electrical and Computer Engineering, DOI 10.1007/978-1-4614-5508-0_6, © The Authors 2013

Cu. Wei et al. demonstrate large current densities in excess of 10^9 A/cm^2 at 250°C for two weeks without measurable deformation or breakdown [112]. In addition, CNTs have large current-carrying capabilities. CNTs can have current densities greater than 10^9 A/cm^2 [58, 69] in contrast to metallic densities of 10^5 A/cm^2 [15]. Bundles of CNTs provide redundancy and fault tolerance, and reduce resistivities. While several studies address the benefits of CNT as interconnect, only one recent study by Naeemi et al. investigates the trade-offs in using either SWCNTs or MWCNTs for on-chip power delivery [79]. To outperform Cu using SWCNTs, a minimum metallic nanotube density of 1 per 2.5 nm^2 is required. A minimum segment length larger than 20 micron is required for MWCNTs to outperform an equivalent Cu power grid. Existing fabrication techniques [29, 118, 124], however, have not been able to produce bundles of CNTs that are purely metallic.

CNTs usually grow in the vertical direction and various techniques have been proposed to fabricate CNT vias. Duesberg et al. demonstrated chemical vapor deposition (CVD) technique to fabricate CNTs at temperature of 800°C [36]. Mizuhisa et al. and Cantoro et al. have extended this technique and were able to grow CNTs at temperatures up to 450°C and 400°C, respectively [24, 81]. All of these techniques fabricate CNTs vertically forming the via. Techniques to grow CNTs in the horizontal direction (parallel to silicon substrate) have not been explored in much detail. Anyuan et al. have shown the growth of CNT bundles in parallel to the surface by inhibiting the growth in other directions [25]. Ural et al. used electric field with high temperature CVD to control the alignment of CNT growth [108]. In this work, we assume that TSVs can be fabricated at lower temperatures. For on-chip power grid, we propose growing bundled CNTs (BCNTs) on a separate silicon die (an interposer) and integrating it with the chip through bonding. An interposer die is a thin layer of silicon used to ease design complexity within the package. The interposer typically contains metal routing and vias. It offers a 20× to 100× increase in wiring density over traditional organic and ceramic packaging [57]. The interposer is placed between the C4 bumps and the chip. The chip connects to the interposer through Cu I/O, solder I/O bonding, or micro bumps [60]. The resistance, capacitance, and inductance of these connectors are minimal because of their aspect ratios. For example, Alam et al. report a capacitance less than 2 fF and a resistance close to 40 mΩ [7].

6.2   Design Setup

We use a 3-D IC, as described in Sect. 4.2.1, consisting of three dies: a quad-core chip-multiprocessor (PROC), a memory (MEM), and an accelerator engine (ACCL). We use the trace selection strategy, described in Sect. 4.2.3, to estimate the power requirements of each functional block in the 3-D IC. We use both the off-chip and the on-chip components of the power delivery network (PDN) from Sect. 4.2.2. We use CNTs for two components of the 3-D PDN: as on-chip power grid and as TSVs. Our four design configurations are shown in Table 6.1. All other design parameters remain the same.

Table 6.1 Design configurations of CNT and Cu based 3-D PDN

Parameter	Value
CNT-Grid-CNT-TSV	CNTs for both on-chip power grid and TSVs
CNT-Grid-Cu-TSV	CNTs for on-chip power grid and Cu is used for TSVs
Cu-Grid-CNT-TSV	Cu-based on-chip power grid and CNTs are used for TSVs
Cu-Grid-Cu-TSV	Cu for both on-chip power grid and TSVs

Table 6.2 Dimensions of a grid element in on-chip power grid

Parameter	Value (μm)
Length	208
Width	104
Height	0.5

Fig. 6.1 Mixed BCNT

Each element in the on-chip power grid is modeled as a resistor and an inductor, connected in series as shown in Fig. 4.4. The length of each grid element is calculated from the die size (1×1 cm) and the on-chip grid granularity. We use a 24×24 on-chip power grid. Table 6.2 gives the physical dimensions of the grid element. We assume a uniform distribution of decoupling capacitances as described in Sect. 4.2.2.

We assume a mixed bundle of CNTs: a mixture of SWCNTs and MWCNTs, including metallic and semi-conducting CNTs with diameter variations as illustrated in Fig. 6.1. We utilize a tool, Carbon Nanotubes Interconnect Analyzer (CNIA),

Table 6.3 Parameters used
to model a CNT bundle

Parameter	Value
Density of CNT (tubes/cm^2)	4.5 E+11
Avg diameter	16 nm
Standard deviation	8 nm
Inner–Outer diameter ratio	0.5
Probability of metallic	0.3
Ambient temperature	60°C

[43, 111], for computing the bundle's electrical and thermal properties. We assume
CNT bundle density of 4.5 E+411 tubes/cm^2, less than the maximum density of
5 E+11 tubes/cm^2 for the selected diameter. Carbon nanotubes can be metallic or
semi conducting depending upon their diameters and chiralities [67]. CNTs have
three types of chiralities: zigzag, armchair, and chiral. Out of these three only
armchair nanotubes are metallic and the other two are mostly semi conducting [33].
We therefore assume the probability of metallic tubes to be 0.3. We assumed an
operating temperature of 60°C, typical of modern processors. A summary of our
parameters are shown in Table 6.3.

The BCNT resistance is dictated by the geometry assumed for the CNT power
grid design. Each resulting BCNT segment length is longer than the mean free path
when the resistance is dependent on the tube length, as observed experimentally
[82, 100]. The contact resistance depends on nanotube diameter and processing
technology. Earlier approaches assumed contact resistance to be a major part of
the total resistance, but laboratory experiments have demonstrated that contact
resistance is relatively small [14]. Li et al. report an overall resistance of 35 Ω for
a 25 μm long MWCNT with an outer diameter of 100 nm [64]. Moreover, Massoud
et al. show that for global level interconnect with large widths, the contact resistance
of the bundle is insignificant [68]. In this work, we assume zero contact resistance.

Kinetic and magnetic effects contribute to CNT inductance. Kinetic inductance
reflects the net sum of kinetic energy on either side of moving electrons in a
nanotube. Earlier experimental observations reported 0.1–4.2 nH/μm [103], while
later results report no kinetic inductance [22]. Wei Wang et al. give the mathematical
details about the relationship of width and kinetic inductance of CNT bundles
[111]. For smaller diameters, kinetic inductance is an order of magnitude larger
than magnetic inductance. Kinetic inductance depends on the number of shells and
number of conduction channels in each shell. Magnetic inductance is produced by
time varying currents and the induced magnetic fields. It is dependent on the current
loop consisting of the signal line and its return paths. Some work assumes that
kinetic inductance dominates magnetic inductance [98], while other work develops
a detailed partial inductance model which assumes return paths at infinity [16].

For our analysis, we chose diameter sizes for our bundle that results in kinetic and
magnetic inductances of similar magnitudes. We varied diameter mean and standard
deviation and computed kinetic inductance. The results are shown in Fig. 6.2. For
all these variations, magnetic inductance remained constant at 1.03 E−10H. This

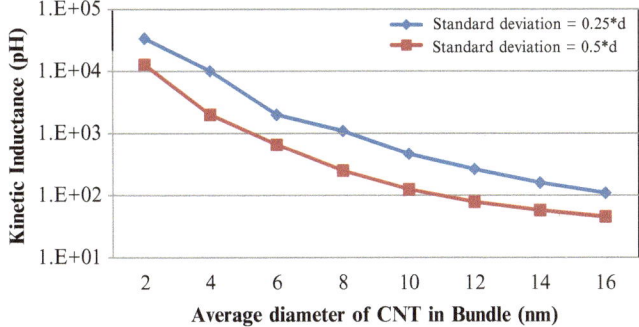

Fig. 6.2 Variations in kinetic inductance with respect to average diameter size (x-axis), plotted for two standard deviation assumptions: 0.25 the diameter, and 0.5 the diameter

Table 6.4 Values of R and L
for each grid element

Technology	R	L
Cu	1.03E−01Ω	8.20E−11H
CNT	3.19E−03Ω	1.48E−10H

shows that kinetic inductance varies exponentially with CNT diameter and as we increase the diameter, kinetic inductance reduces until it becomes comparable to the magnetic one. We choose larger diameters with higher standard deviations for our PDN design.

To compute the R & L values of Cu grid elements, we use the Predictive Technology Model (PTM) [3] with resistivity $2.2\,\mu\Omega$-cm and thermal co-efficient of $3.93\,\mathrm{E}{-}03/^\circ\mathrm{C}$. Values are shown in Table 6.4.

6.3   Substrate Area Dedicated to TSV for Power Delivery

In this experiment, our goal is to evaluate CNTs to reduce the chip area dedicated to TSVs delivering power in a 3-D IC. We explore the TSV area penalty for each of the four configurations. To find the minimum number of TSVs dedicated for power delivery, we use the Improve Worst Violation (IWV) algorithm described in Sect. 5.3.4. We run this algorithm for an IR budget of 12% whereas VDD is assumed to be 1V.

The total number of TSVs for each benchmark are given in Table 6.5 (rows 1–4). We use the maximum number of TSVs at each TSV location across all the benchmarks to determine the required number of TSVs. The total number of TSVs and the distribution of TSVs over the floorplan for each 3-D IC configuration are given in Table 6.5 (row 5) and Fig. 6.3, respectively. We conclude the following:

• Using CNTs for on-chip power grid and as TSVs requires relatively few TSVs to meet the power budget and hence tremendously reduces the area penalty. CNT-Grid-CNT-TSV configuration requires the smallest number of TSVs among the

Table 6.5 Number of TSVs required for each configuration

Benchmark	CNT grid		Cu grid	
	CNT TSV	Cu TSV	CNT TSV	Cu TSV
apsi	477	684	871	1,363
bzip	523	716	1,134	1,866
equake	492	589	598	756
mcf	525	706	989	1,564
Across all benchmarks	553	736	1,145	1,887

four proposed configurations. The total number of TSVs for CNT-Grid-CNT-TSV are $0.75\times$, $0.48\times$, and $0.29\times$ for CNT-Grid-Cu-TSV, Cu-Grid-CNT-TSV, and Cu-Grid-Cu-TSV, respectively.

- Considering the Cu-Grid-Cu-TSV as the base case, TSV area is reduced to $0.38\times$ by changing the grid to CNT. Similarly, we get $0.60\times$ reduction when we fabricate TSVs using CNT. This suggests that CNTs as the on-chip power grid have more impact than CNTs as TSVs.
- Although CNTs as on-chip power grid have a large impact, CNT TSVs still result in significant reduction in area penalty. Our results show that for a CNT grid, CNT TSVs reduce the area penalty by $0.75\times$.
- Results presented in Fig. 6.3 show that at most three TSVs are required at any grid point for the CNT grid, whereas 14 are required for the Cu-grid.

6.4 IR Analysis

Once we calculate the optimal number of TSVs for each configuration, we can evaluate IR drops for the four configurations. We model the power grid as resistive elements. We use the compressed traces from four SPEC benchmarks (apsi, bzip, equake, and mcf) as described in Sect. 4.2.3. We study the impact of using CNTs relative to using Cu in a 3-D PDN. For this purpose, we consider the TSV placement for CNT-Grid-CNT-TSV configuration as a baseline and use it for the three other configurations. All design parameters other than TSV placement remain the same. We perform IR analysis for each benchmark and report the maximum voltage drop, the average voltage drop, and the standard deviation in Table 6.6. We conclude the following:

- For the CNT grid, there is little difference between using CNT and Cu TSVs for all dies. The CNT power grid dominates the overall performance of the PDN and using either Cu or CNT TSVs has little impact.
- The impact of using CNT TSVs is more pronounced when using the Cu-based grid. For the PROC die, there is a 32% and 15% improvement in the maximum and average IR drops, respectively. CNT TSVs reduce the standard deviation by 24%, which indicates a smaller voltage variation across the PROC die.

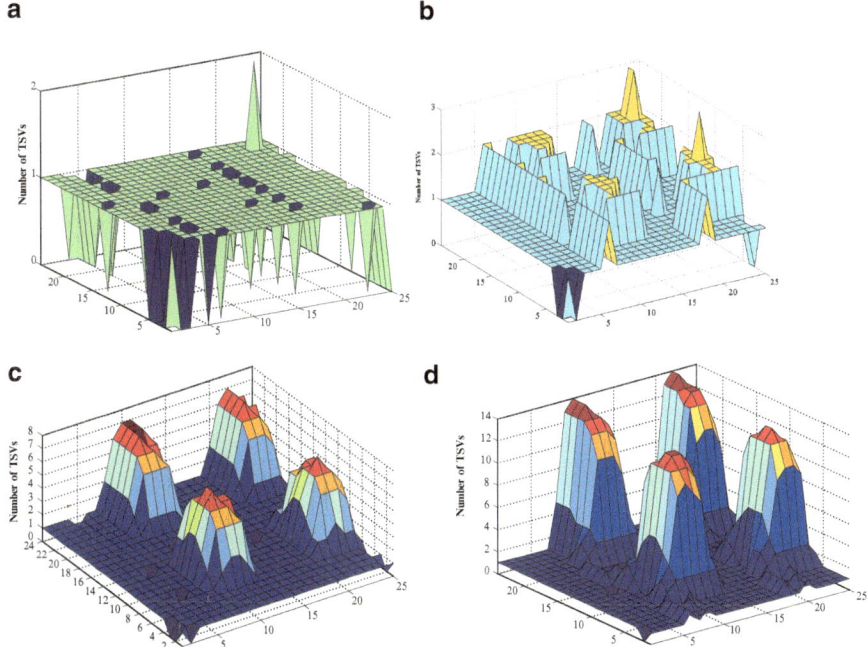

Fig. 6.3 Substrate area dedicated to TSVs for power delivery in the 3-D IC. (**a**) CNT-Grid-CNT-TSV, (**b**) CNT-Grid-Cu-TSV, (**c**) Cu-Grid-CNT-TSV, (**d**) Cu-Grid-Cu-TSV

- The die placement within a 3-D stack affects power quality. ACCL, which is closest to the package, has the least improvement because it does not have any TSVs. The CNT grid improves PROC and MEM, but not ACCL. The improvement due to CNT TSVs is more pronounced for PROC and only when using the Cu grid.
- The CNT power grid performs better than the Cu power grid, showing an improvement for PROC of 98% when using Cu for both the on-chip grid and for TSVs. With Cu TSVs, the CNT grid has a PROC improvement of 66% over using a Cu grid.

6.5 Ldi/dt Analysis

The strategy to evaluate Ldi/dt performance for CNTs is similar to the one we used in IR analysis, except that we additionally consider inductive and capacitive components. We use the optimal number of TSVs for the CNT-Grid-CNT-TSV configuration. The results are presented in Table 6.7. We conclude the following:

Table 6.6 IR voltage drop for different 3-D configurations (normalized to CNT-Grid-CNT-TSV)

	PROC				MEM				ACCL			
	CNT grid		Cu grid		CNT grid		Cu grid		CNT grid		Cu grid	
	CNT TSV	Cu TSV	CNT TSV	Cu TSV	CNT TSV	Cu TSV	CNT TSV	Cu TSV	CNT TSV	Cu TSV	CNT TSV	Cu TSV
apsi												
Maximum voltage drop	1.000	0.994	1.660	1.982	1.000	0.996	1.304	1.469	1.000	1.003	1.072	1.070
Average voltage drop	1.000	1.000	1.240	1.391	1.000	1.000	1.119	1.205	1.000	1.000	1.040	1.040
Standard deviation	1.000	1.000	1.365	1.595	1.000	1.000	1.193	1.320	1.000	1.000	1.000	0.999
bzip												
Maximum voltage drop	1.000	0.994	1.657	1.975	1.000	0.996	1.299	1.462	1.000	1.003	1.071	1.070
Average voltage drop	1.000	1.000	1.246	1.400	1.000	1.000	1.120	1.207	1.000	1.000	1.038	1.038
Standard deviation	1.000	1.000	1.378	1.613	1.000	1.000	1.194	6.961	1.000	1.000	1.003	1.002
equake												
Maximum voltage drop	1.000	0.996	1.641	1.979	1.000	0.997	1.270	1.428	1.000	1.002	1.063	1.063
Average voltage drop	1.000	1.000	1.255	1.415	1.000	1.000	1.124	1.215	1.000	1.000	1.035	1.035
Standard deviation	1.000	1.000	1.373	1.605	1.000	1.000	1.190	1.315	1.000	1.000	1.008	1.007
mcf												
Maximum voltage drop	1.000	0.995	1.631	1.940	1.000	0.996	1.290	1.449	1.000	1.003	1.068	1.067
Average voltage drop	1.000	1.000	1.251	1.409	1.000	1.000	1.122	1.211	1.000	1.000	1.037	1.037
Standard deviation	1.000	1.000	1.394	1.637	1.000	1.000	1.196	1.327	1.000	1.000	1.000	0.999

Table 6.7 Ldi/dt voltage drop for different 3-D configurations (normalized to CNT-Grid-CNT-TSV)

	PROC				MEM				ACCL			
	CNT grid		Cu Grid		CNT grid		Cu grid		CNT grid		Cu grid	
	CNT TSV	Cu TSV	CNT TSV	Cu TSV	CNT TSV	Cu TSV	CNT TSV	Cu TSV	CNT TSV	Cu TSV	CNT TSV	Cu TSV
apsi												
Maximum voltage drop	1.000	1.051	1.272	1.475	1.000	1.035	0.966	1.060	1.000	0.952	0.906	0.872
Average voltage drop	1.000	1.123	1.058	1.185	1.000	1.079	1.036	1.118	1.000	0.996	1.016	1.014
Standard deviation	1.000	0.998	0.971	1.029	1.000	0.973	0.932	0.939	1.000	0.947	0.946	0.904
bzip												
Maximum voltage drop	1.000	0.930	0.984	1.096	1.000	0.922	0.702	0.738	1.000	0.972	0.923	0.899
Average voltage drop	1.000	1.120	1.050	1.180	1.000	1.077	1.031	1.114	1.000	0.995	1.014	1.011
Standard deviation	1.000	0.957	0.940	0.960	1.000	0.944	0.910	0.881	1.000	0.929	0.930	0.874
equake												
Maximum voltage drop	1.000	0.954	1.068	1.188	1.000	0.879	0.768	0.766	1.000	0.931	0.871	0.846
Average voltage drop	1.000	1.119	1.052	1.177	1.000	1.076	1.036	1.117	1.000	0.994	1.012	1.009
Standard deviation	1.000	1.007	0.998	1.032	1.000	0.988	0.963	0.966	1.000	0.969	0.975	0.951
mcf												
Maximum voltage drop	1.000	1.000	1.015	1.139	1.000	0.922	0.856	0.908	1.000	0.985	0.993	0.975
Average voltage drop	1.000	1.119	1.050	1.176	1.000	1.078	1.037	1.120	1.000	0.997	1.023	1.021
Standard deviation	1.000	0.989	0.948	0.994	1.000	0.965	0.905	0.897	1.000	0.944	0.923	0.885

- CNT TSVs result in better performance for both grid types when compared to Cu. For the CNT-grid, the improvement in the maximum and average voltage drops in the PROC die are 5% and 12% respectively. For the Cu-grid, these improvements are 20% and 13%. These results show that even with fewer TSVs, CNT TSVs outperform Cu TSVs.
- The transient behavior of Cu is better than that of CNTs for ACCL as it is connected directly to the package using C4 bumps and there was no change in the off-chip PDN or the number of C4 bumps. Decreasing the number of TSVs results in reducing sharing across dies and isolating ACCL from the rest of the 3-D IC. This in turn results in local improvements for ACCL.
- Although the CNT grid performs better than the Cu grid, the improvement in Ldi/dt is not as significant as the one obtained for IR due to CNT inductance. For example, CNTs result in an improvement as high as 98% and 66% for maximum and average IR drops for the PROC die. The same numbers for Ldi/dt analysis are 47% and 5%.

6.6 Summary

We explored in this chapter the feasibility of using CNTs as TSVs and as on-chip power grid. Our results show that the area required for power delivery can be significantly reduced when using CNT over Cu TSVs. Using CNTs for TSVs only and for both TSVs and on-chip power grid reduced the area requirement by 40% and 71%, respectively, over an equivalent Cu implementation. We performed IR and Ldi/dt analysis for different design configurations with CNT-optimal TSV placements. Using CNTs either for the on-chip power grid, or for TSVs or for both improved power quality. For the same optimal substrate area dedicated to Cu-based power delivery, CNTs can improve the average voltage drop by 40% for IR analysis and by 18% for Ldi/dt analysis.

Chapter 7
Conclusions and Future Directions

This book investigates challenges associated with TSVs in 3-D ICs. This work is innovative as it explores design and technology challenges, evaluates potential benefits, and proposes novel solutions. The results presented in this book advance the understanding of TSV-based 3-D IC design and guide designers in selecting design and technology parameters.

7.1 Summary

This book offers original contributions in several key areas. First, a new technique in mitigating TSV-induced substrate noise is presented. We develop a realistic framework to study TSV-induced substrate noise. This framework uses a finite-element three-dimensional tool to extract lumped parasitics of the design setups. We use this framework to evaluate the performance of different noise mitigation techniques in terms of noise isolation and substrate area penalty. We report that shielding TSVs from devices using a dielectric liner is not sufficient to mitigate noise; substrate grounding is required. For this purpose, we propose a new technique, the GND plug, to effectively ground the substrate. We show that the GND plug is a superior technology in reducing substrate noise while utilizing small area and minimizing the keep out zone. For a $2\,\mu m$ TSV in a $20\,\mu m$ thick substrate, four GND plugs of diameter $0.5\,\mu m$ fabricated around the TSV at a distance of $1\,\mu m$ result in substrate noise less than 8 % of the input voltage. We analyze the size and placement considerations of GND plugs to maximize noise isolation and to minimize area penalty. The GND plug is a better technique than increasing the thickness of the dielectric liner or using a backside GND plane. Compared to using a $1.5\,\mu m$ thick liner, a GND plug reduces substrate noise by $4.3\times$ and reduces the area by $0.67\times$. Compared to a backside GND plane, the GND plug technique is simple and results in $4.3\times$ performance improvement.

N. Khan and S. Hassoun, *Designing TSVs for 3D Integrated Circuits*, SpringerBriefs in Electrical and Computer Engineering, DOI 10.1007/978-1-4614-5508-0_7, © The Authors 2013

Second, we develop a trace selection strategy to speed IR power delivery analysis without compromising quality. This strategy utilizes an architectural model of the IC to extract block-level current traces and picks only representative ones. We use this approach on 10 million traces from four benchmarks (apsi, bzip, equake, and mcf). We reduce the number of traces to 1,428.

Third, we perform the first architectural-level power delivery analysis for 3-D ICs utilizing realistic workloads. Within this study, we analyze the impact of TSV size and spacing to assess the trade offs between TSV area and PDN performance. We observe that PDN performance saturates for a TSV size of 25 μm and a TSV granularity of 32×32. Increasing the granularity of C4 bumps significantly improves PDN performance. In addition, we show that coaxial TSVs improve PDN performance by providing additional decoupling capacitance, and that coaxial TSVs reduce interconnect and device blockage by overlaying power/signal routing within a single coaxial TSV.

Fourth, we investigate the problem of PDN design at early design stages to estimate required TSV area, allowing designers to arrive at total area and cost estimates for 3-D implementations. To compare the performance of the TSV-estimation algorithms, we use a practical design consisting of three dies. Starting with a minimal TSV area estimate and incrementing iteratively produces the smallest number of TSVs compared to starting with a larger estimate and decreasing iteratively. This technique, IMPROVE WORST VIOLATION (IWV), resulted in a 40 % improvement in our evaluation framework.

Finally, we evaluate the use of carbon nanotubes for power delivery in 3-D ICs for the on-chip power grid and for TSVs. We use a practical CNT model to calculate RLC values of CNT bundles. We use the IWV algorithm to estimate the total number of TSVs required to deliver power within the noise budget. Our results show that using CNTs as TSVs provides at least 40 % reduction is TSV area when compared to a Cu TSV implementation. A reduction of 71 % is achieved when CNTs are used for both on-chip power grid and TSVs. For the same substrate area dedicated to TSVs for power delivery, CNTs in a 3-D PDN perform better than Cu. CNTs are capable of improving the average voltage drop by 40 % and 18 % for IR and Ldi/dt analysis, respectively.

7.2 Future Research Directions

Stacking of 3-D ICs is a new and emerging technology. This book covers a few aspects of TSV-based 3-D IC design. A large number of design and technology challenges still need to be met to bring 3-D IC technology to mainstream VLSI design. A few possible extensions and some future challenges include:

- The GND plug technique proposed in this book was evaluated using only simulations. Test chips with different configurations of TSVs and GND plugs are the next logical step in evaluating this technology. More complex coupling scenarios must be investigated to assess the impact of coupling.

- Analytical models and layout extraction tools are required to efficiently analyze device-to-TSV and TSV-to-TSV coupling.
- TSVs create thermal stress, which impacts the performance of neighboring devices. Analysis tools to quantify the impact of thermal stress on device performance and techniques to reduce this impact are required. Both TSV-induced noise and TSV-induced stress dictate the size of keep out zone for devices. Further analysis of these two phenomenon needs to be performed to create new design rules for devices in 3-D ICs.
- Thermal management is a challenge in 3-D ICs. TSVs are proposed to extract heat from dies away from the heat spreader. Detailed analysis that consider dielectric liner and practical TSV placement are needed. Coupled thermal and power delivery analysis is are required to estimate the substrate area needed for TSVs.

3-D integration is poised to deliver improved IC performance and heterogeneous die integration. Solving 3-D design challenges will drive this technology, thus advancing a wide range of medical, consumer, and high performance electronic systems.

References

1. Ansoft - Q3D Extractor. http://www.ansoft.com/products/si/q3d_extractor/, URL http://www.ansoft.com/products/si/q3d_extractor/
2. MITLL Low-Power FDSOI CMOS Process. http://www.ece.umd.edu/~dilli/research/layout/MITLL_3D_2006/3D_PDK2.3/doc/ApplicationNotes2006-1.pdf, URL http://www.ece.umd.edu/~dilli/research/layout/MITLL_3D_2006/3D_PDK2.3/doc/ApplicationNotes2006-1.pdf
3. Predictive technology model (PTM). http://www.eas.asu.edu/~ptm/
4. Redistributed Chip Packaging (RCP) Technology. http://www.freescale.com/webapp/sps/site/overview.jsp?code=ASIC_LV3_PACKAGING_RCP. URL http://www.freescale.com/webapp/sps/site/overview.jsp?code=ASIC_LV3_PACKAGING_RCP
5. (2009) International Technology Roadmap for Semiconductors. http://wwwitrsnet/Links/2009ITRS/Home2009htm URL http://www.itrs.net/
6. Afzali-Kusha A, Nagata M, Verghese N, Allstot D (2006) Substrate noise coupling in SoC design: modeling, avoidance, and validation. Proc IEEE 94(12):2109–2138
7. Alam SM, Jones RE, Rauf S, Chatterjee R (2007) Inter-Strata connection characteristics and signal transmission in three-dimensional (3D) integration technology. In: 8th international symposium on quality electronic design, pp 580–585
8. Andry P, Tsang C, Sprogis E, Patel C, Wright S, Webb B, Buchwalter L, Manzer D, Horton R, Polastre R, Knickerbocker J (2006) A CMOS-compatible process for fabricating electrical through-vias in silicon. In: Electronic components and technology conference, pp 831–837
9. Andry PS, Tsang CK, Webb BC, Sprogis EJ, Wright SL, Dang B, Manzer DG (2008) Fabrication and characterization of robust through-silicon vias for silicon-carrier applications. IBM J Res Develop 52(6):571–581
10. Bakir M, King C, Sekar D, Thacker H, Dang B, Huang G, Naeemi A, Meindl J (2008) 3D Heterogeneous integrated systems: liquid cooling, power delivery, and implementation. In: IEEE custom integrated circuits conference, pp 663–70
11. Bamal M, List S, Stucchi M, Verhulst A, Hove MV, Cartuyvels R, Beyer G, Maex K (2006) Performance comparison of interconnect technology and architecture options for deep submicron technology nodes. In: International interconnect technology conference, pp 202–204
12. Banerjee K, Srivastava N (2006) Are carbon nanotbues the future of VLSI interconnects? In: 43rd IEEE/ACM annual conference on design automation, pp 809–814
13. Banerjee K, Im S, Srivastava N (2005) Interconnect modeling and analysis in the nanometer era: Cu and beyond. In: 22nd advanced metallization conference
14. Banerjee K, Li H, Srivastava N (2008) Current status and future perspectives of carbon nanotube interconnects. In: 8th IEEE conference on nanotechnology, pp 432–436

N. Khan and S. Hassoun, *Designing TSVs for 3D Integrated Circuits*, SpringerBriefs in Electrical and Computer Engineering, DOI 10.1007/978-1-4614-5508-0,
© The Authors 2013

15. Baughman RH, Zakhidov AA, De Heer WA (2002) Carbon nanotubes – the route toward applications. Science 297(5582):787–792
16. Beattie M, Pileggi L (2001) Inductance 101: modeling and extraction. In: 38th design automation conference, pp 323–328
17. Beica R, Siblerud P, Sharbono C, Bernt M (2008) Advanced metallization for 3D integration. In: 10th electronics packaging technology conference, pp 212–218
18. Beyne E (2008) Solving technical and economical barriers to the adoption of Through-Si-Via 3D integration technologies. In: 10th electronics packaging technology conference, pp 29–34
19. Bhattacharya U, Wang Y, Hamzaoglu F, Ng Y, Wei L, Chen Z, Rohlman J, Young I, Zhang K (2008) 45nm SRAM technology development and technology lead vehicle. Intel Tech J 12(02)
20. Borkar S (2009) Design perspectives on 22nm CMOS and beyond. In: 46th ACM/IEEE design automation conference, pp 93–94
21. Brooks D, Tiwari V, Martonosi M (2000) Wattch: a framework for architectural-level power analysis and optimizations. In: 27th international symposium on computer architecture, pp 83–94
22. Burke P (2002) Luttinger liquid theory as a model of the gigahertz electrical properties of carbon nanotubes. IEEE Trans Nanotechnol 1(3):129–144
23. Burns J, Aull B, Chen C, Chen C, Keast C, Knecht J, Suntharalingam V, Warner K, Wyatt P, Yost D (2006) A wafer-scale 3-D circuit integration technology. IEEE Trans Electron Dev 53(10):2507–2516
24. Cantoro M, Hofmann S, Pisana S, Scardaci V, Parvez A, Ducati C, Ferrari A, Blackburn A, Wang K, Robertson J (2006) Catalytic chemical vapor deposition of single-wall carbon nanotubes at low temperatures. Nano Lett 6(6):1107–1112
25. Cao A, Baskaran R, Frederick M, Turner K, Ajayan P, Ramanath G (2003) Direction-selective and length-tunable in-plane growth of carbon nanotubes. Adv Mater 15(13):1105–1109
26. Chang M (2007) Foundry future: challenges in the 21st century. In: IEEE international solid-state circuits conference, pp 18–23
27. Chen D, Chiou W, Chen M, Wang T, Ching K, Tu H, Wu W, Yu C, Yang K, Chang H, Tseng M, Hsiao C, Lu Y, Hu H, Lin Y, Hsu C, Shue W, Yu C (2009) Enabling 3D-IC foundry technologies for 28 nm node and beyond: through-silicon-via integration with high throughput die-to-wafer stacking. In: IEEE international electron devices meeting, pp 1–4
28. Chen T, Chen CC (2001) Efficient large-scale power grid analysis based on precondi-tioned krylov-subspace iterative methods. In: IEEE/ACM design automation conference, pp 559–562
29. Cheung C, Kurtz A, Park H, Lieber C (2002) Diameter-controlled synthesis of carbon nanotubes. J Phys Chem B 106(10):2429–2433
30. Cho J, Shim J, Song E, Pak JS, Lee J, Lee H, Park K, Kim J (2009) Active circuit to through silicon via (TSV) noise coupling. In: IEEE 18th conference on electrical performance of electronic packaging and systems, pp 97–100
31. Clement F (2001) Substrate noise coupling analysis in mixed-signal ICs. Presentation from the workshop on substrate-noise coupling in mixed-signal ICs, IMEC, Leuven, Belgium, September 2001
32. Cong J, Zhang Y (2005) Thermal-driven multilevel routing for 3-D ICs. In: Asia and South Pacific design automation conference, pp 121–126
33. Datta S (2005) Quantum transport: atom to transistor, 2nd edn. Cambridge University Press, Cambridge
34. Davis W, Wilson J, Mick S, Xu J, Hua H, Mineo C, Sule A, Steer M, Franzon P (2005) Demystifying 3D ICs: The Pros and Cons of going vertical. IEEE Des Test Comput 22(6):498–510
35. Denda S (2007) Process examination of through silicon via technologies. In: 6th international conference on polymers and adhesives in microelectronics and photonics, pp 149–152

36. Duesberg GS, Graham AP, Kreupl F, Liebau M, Seidel R, Unger E, Hoenlein W (2004) Ways towards the scaleable integration of carbon nanotubes into silicon based technology. Diam Relat Mater 13(2):354–361
37. Early J (1960) Speed, power and component density in multielement high-speed logic systems. In: IEEE international solid-state circuits conference, vol III, pp 78–79
38. Garrou P, Bower C, Ramm P (2008) Handbook of 3D integration: technology and applications of 3D integrated circuits. Wiley-VCH, Weinheim
39. Goering R (2009) A qualcomm perspective on 3D ICs. http://wwwcadencecom/community/blogs/ii/archive/2009/04/20/a-qualcomm-perspective-on-3d-icsaspx
40. Golshani1 N, Derakhshandeh1 J, Ishihara1 R, Beenakker C, Robertson2 M, Morrison T (2010) Monolithic 3D integration of SRAM and image sensor using two layers of single grain silicon. In: IEEE international conference on 3D system integration, pp 1–7
41. Goplen B, Sapatnekar S (2006) Placement of thermal vias in 3-D ICs using various thermal objectives. IEEE Trans Computer Aided Des Integrated Circ Syst 25(4):692–709
42. Gupta M, Oatley J, Joseph R, Wei G, Brooks D (2007) Understanding voltage variations in chip multiprocessors using a distributed power-delivery network. In: Design, automation test in Europe, pp 1–6
43. Haruehanroengra S, Wang W (2007) Analyzing conductance of mixed carbon-nanotube bundles for interconnect applications. IEEE Electron Dev Lett 28(8):756–759
44. Ho SW, Rao VS, Khan QKN, Yoon SU, Kripesh V (2006) Development of coaxial shield via in silicon carrier for high frequency application. In: 8th electronics packaging technology conference, pp 825–830
45. Ho SW, Yoon SW, Zhou Q, Pasad K, Kripesh V, Lau J (2008) High RF performance TSV silicon carrier for high frequency application. In: 58th electronic components and technology conference, pp 1946–1952
46. Huang G, Bakir M, Naeemi A, Chen H, Meindl J (2007) Power delivery for 3D chip stacks: physical modeling and design implication. In: IEEE electrical performance of electronic packaging, pp 205–208
47. Ishikuro H, Miura N, Kuroda T (2007) Wideband inductive-coupling interface for high-performance portable system. In: IEEE custom integrated circuits conference, pp 13–20
48. Jain P, Kim T, Keane J, Kim CH (2008) A multi-story power delivery technique for 3D integrated circuits. In: 13th international symposium on low power electronics and design, pp 57–62
49. Jang DM, Ryu C, Lee KY, Cho BH, Kim J, Oh TS, Lee WJ, Yu J (2007) Development and evaluation of 3-D SiP with vertically interconnected through silicon vias (TSV). In: Electronic components and technology conference, pp 847–852
50. Joyner J, Venkatesan R, Zarkesh-Ha P, Davis J, Meindl J (2001) Impact of three-dimensional architectures on interconnects in gigascale integration. IEEE Trans Very Large Scale Integ (VLSI) Syst 9(6):922–928
51. Keigler A, O'Donnell K, Liu Z, Wu B, Trezza J (2007) Enabling 3-D design. Semicond Int 30(9):36–44
52. Khan NH, Alam SM, Hassoun S (2011) Power delivery design for 3-D ICs using different through-silicon via (TSV) technologies. IEEE Trans VLSI Syst 19(4):647–658
53. Khan N, Alam S, Hassoun S (2009) Through-silicon via (TSV)-induced noise characterization and noise mitigation using coaxial TSVs. In: IEEE international conference on 3D system integration, pp 1–7
54. Kikuchi H, Yamada Y, Ali AM, Liang J, Fukushima T, Tanaka T, Koyanagi M (2008) Tungsten through-silicon via technology for three-dimensional LSIs. Jpn J Appl Phys 47:2801–2805
55. Kim B, Sharbono C, Ritzdorf T, Schmauch D (2006) Factors affecting copper filling process within high aspect ratio deep vias for 3D chip stacking. In: 56th electronic components and technology conference, pp 838–843

56. King C, Sekar D, Bakir M, Dang B, Pikarsky J, Meindl J (2008) 3D Stacking of chips with electrical and microfluidic I/O interconnects. In: Electronic components and technology conference, pp 1–7

57. Knickerbocker J, Patel C, Andry P, Tsang C, Buchwalter L, Sprogis E, Gan H, Horton R, Polastre R, Wright S, Cotte J (2006) 3-D silicon integration and silicon packaging technology using silicon through-vias. IEEE J Solid State Circ 41(8):1718–1725

58. Kreupl F, Graham AP, Liebau M, Duesberg GS, Seidel R, Unger E (2004) Carbon nanotubes for interconnect applications. In: International electron devices meeting, pp 683–686

59. Kuo WS, Wang M, Chen E, Lai JY, Wang YP (2008) Thermal investigations of 3D FCBGA packages with TSV technology. In: 3rd international microsystems, packaging, assembly circuits technology conference, pp 251–254

60. Kurita Y, Soejima K, Kikuchi K, Takahashi M, Tago M, Koike M, Shibuya K, Yamamichi S, Kawano M (2006) A novel "SMAFTI" package for inter-chip wide-band data transfer. In: 56th electronic components and technology conference, pp 289–297

61. Laviron C, Dunne B, Lapras V, Galbiati P, Henry D, Toia F, Moreau S, Anciant R, Brunet-Manquat C, Sillon N (2009) Via first approach optimisation for through silicon via applications. In: 59th electronic components and technology conference, pp 14–19

62. Lee Y, Goel R, Lim SK (2009) Multi-functional interconnect co-optimization for fast and reliable 3D stacked ICs. In: IEEE/ACM international conference on computer-aided design, pp 645–51

63. Lee Y, Yoon JK, Gang H, Bakir M, Joshi Y, Fedorov A, Sung KL (2009) Co-design of signal, power, and thermal distribution networks for 3D ICs. In: Design, automation and test in Europe, pp 610–615

64. Li H, Lu W, Li J, Bai X, Gu C (2005) Multichannel ballistic transport in multiwall carbon nanotubes. Phys Rev Lett 95(8):86,601–86,601

65. Loh GH (2008) 3D-stacked memory architectures for multi-core processors. In: 35th international symposium on computer architecture, pp 453–464

66. Loh GH, Xie Y, Black B (2007) Processor design in 3D die-stacking technologies. Micro IEEE 27(3):31–48

67. Loiseau A, Launois P, Petit P, Roche S, Salvetat J (2006) Understanding carbon nanotubes: from basics to applications. Springer, New York

68. Massoud Y, Nieuwoudt A (2006) Modeling and design challenges and solutions for carbon nanotube-based interconnect in future high performance integrated circuits. ACM J Emerg Tech Comput Syst 2(3):155–196

69. McEuen P, Park JY (2004) Electron transport in single-walled carbon nanotubes. MRS Bull 29(4):272–275

70. Meindl J (2003) Beyond moore's law: the interconnect era. Comput Sci Eng 5(1):20–24

71. Miao M, Jin Y, Liao H, Zhao L, Zhu Y, Sun X, Guo Y (2009) Research on deep RIE-based through-si-via micromachining for 3-D system-in-package integration. In: 4th IEEE international conference on nano/micro engineered and molecular systems, pp 90–93

72. Minz JR, Lim SK, Koh C (2005) 3D module placement for congestion and power noise reduction. In: Proceedings of the 15th ACM Great Lakes symposium on VLSI, pp 458–461

73. Mofrad MRT, Derakhshandeh J, Ishihara R, Baiano A, van der Cingel J, Beenakker K (2009) Stacking of single-grain thin-film transistors, Japanese journal of applied physics 48:03B015-03B015-4

74. Moore G (2003) No exponential is forever: but "Forever" can be delayed! In: IEEE international solid-state circuits conference, pp 20–23

75. Moore GE (1965) Cramming more components onto integrated circuits. Electronics 38(8):114–117

76. Morrow P, Kobrinsky M, Ramanathan S, Park C, Harmes M, Ramachandrarao V, mog Park H, Kloster G, List S, Kim S (2005) Wafer-level 3D interconnects via Cu bonding. In: Advanced metallization conference, pp 125–30

77. Motoyoshi M (2009) Through-silicon via (TSV). Proc IEEE 97(1):43–48

78. Naeemi A, Meindl JD (2009) Carbon nanotube interconnects. Ann Rev Mater Res 39:255–275
79. Naeemi A, Huang G, Meindl JD (2007) Performance modeling for carbon nanotube interconnects in on-chip power distribution. In: Electronic components and technology conference, pp 420–428
80. Nagarajan R, Ebin L, Dayong L, Seng SC, Prasad K, Balasubramanian N (2006) Development of a novel deep silicon tapered via etch process for through-silicon interconnection in 3-D integrated systems. In: 56th electronic components and technology conference, pp 383–387
81. Nihei M, Kondo D, Kawabata A, Sato S, Shioya H, Sakaue M, Iwai T, Ohfuti M, Awano Y (2005) Low-resistance multi-walled carbon nanotube vias with parallel channel conduction of inner shells. In: IEEE 2005 international interconnect technology conference, pp 234–236
82. Park JY, Rosenblatt S, Yaish Y, Sazonova V, Ustunel H, Braig S, Arias T, Brouwer P, McEuen P (2004) Electron-phonon scattering in metallic single-walled carbon nanotubes. Nano Lett 4(3):517–520
83. Patel CS (2006) Silicon carrier for computer systems. In: 43rd ACM/IEEE design automation conference, pp 857–862
84. Patti R (2006) Three-dimensional integrated circuits and the future of system-on-chip designs. Proc IEEE 94(6):1214–1224
85. der Plas GV, Limaye P, Mercha A, Oprins H, Torregiani C, Thijs S, Linten D, Stucchi M, Guruprasad K, Velenis D, Shinichi D, Cherman V, Vandevelde B, Simons V, Wolf ID, Labie R, Perry D, Bronckers S, Minas N, Cupac M, Ruythooren W, Olmen JV, Phommahaxay A, de Potter de ten Broeck M, Opdebeeck A, Rakowski M, Wachter BD, Dehan M, Nelis M, Agarwal R, Dehaene W, Travaly Y, Marchal P, Beyne E (2010) Design issues and considerations for low-cost 3D TSV IC technology. In: 2010 IEEE international solid-state circuits conference, pp 148–149
86. Pozder S, Lu J, Kwon Y, Zollner S, Yu J, McMahon J, Cale T, Yu K, Gutmann R (2004) Back-end compatibility of bonding and thinning processes for a wafer-level 3D interconnect technology platform. In: IEEE international interconnect technology conference, pp 102–106
87. Rabaey JM, Chandrakasan AP, Nikoli B (2002) Digital integrated circuits: a design perspective. Prentice hall, New Jersey
88. Rahman A, Trezza J, New B, Trimberger S (2006) Die stacking technology for terabit chip-to-chip communications. In: IEEE custom integrated circuits conference, pp 587–590
89. Ramaswami S (2010) Process equipment readiness for through-silicon via technologies. Solid State Tech 53(8):16–17
90. Rousseau M, Rozeau O, Cibrario G, Le Carval G, Jaud M.-A, Leduc P, Farcy A, Marty A (2008) Through-silicon via based 3D IC technology: Electrostatic simulations for design methodology. In: IMAPS device packaging conference, Phoenix, AZ
91. Rousseau M, Jaud M, Leduc P, Farcy A, Marty A (2009) Impact of substrate coupling induced by 3D-IC architecture on advanced CMOS technology. In: microelectronics and packaging conference, pp 1–5
92. Schrom G, Liu D, Pichler C, Svensson C, Selberherr S (1994) Analysis of ultra-low-power CMOS with process and device simulation. In: 24th European solid state device research conference, pp 679–682
93. Schulz M (1999) The End of the Road for Silicon. Nature 399(6738):729–730
94. Selvanayagam C, Lau J, Zhang X, Seah S, Vaidyanathan K, Chai T (2009) Nonlinear thermal stress/strain analyses of copper filled TSV (through silicon via) and their flip-chip microbumps. IEEE Trans Adv Packag 32(4):720–728
95. Selvanayagam C, Zhang X, Rajoo R, Pinjala D (2010) Modelling stress in silicon with TSVs and its effect on mobility. In: 11th electronics packaging technology conference, pp 612–618
96. Singer P (2008) Through-silicon vias: ready for volume manufacturing? Semicond Int 31(3):22–26
97. Sparks TG, Alam SM, Chatterjee R, Rauf S (2006) Method of forming a through-substrate via. U.S. patent appl. 20080113505

98. Srivastava N, Banerjee K (2005) Performance analysis of carbon nanotube interconnects for VLSI applications. In: International conference on computer aided design, pp 383–390

99. Srivastava N, Joshi R, Banerjee K (2005) Carbon nanotube interconnects: implications for performance, power dissipation and thermal management. In: International electron devices meeting, pp 249–252

100. Stahl H, Appenzeller J, Martel R, Avouris P, Lengeler B (2000) Intertube coupling in ropes of single-wall carbon nanotubes. Phys Rev Lett 85(24):5186–5189

101. Sun X, Ji M, Ma S, Zhu Y, Kang W, Miao M, Jin Y (2010) Electrical characterization of sidewall insulation layer of TSV. In: 11th international conference on electronic packaging technology & high density packaging, pp 77–80

102. Tang Z (2010) Efficient design practices for thermal management of a TSV based 3D IC system. In: 19th international symposium on physical design, pp 59–59

103. Tarkiainen R, Ahlskog M, Penttilä J, Roschier L, Hakonen P, Paalanen M, Sonin E (2001) Multiwalled carbon nanotube: luttinger versus fermi liquid. Phys Rev B 64(19):195,412–195,415

104. Tezcan D, Pham N, Majeed B, Moor PD, Ruythooren W, Baert K (2007) Sloped through wafer vias for 3D wafer level packaging. In: 57th electronic components and technology conference, pp 643–647

105. Thompson S, Chau R, Ghani T, Mistry K, Tyagi S, Bohr M (2005) In search of forever, continued transistor scaling one new material at a time. IEEE Trans Semicond Manuf 18(1):26–36

106. Topol A, Tulipe DL, Shi L, Alam S, Frank D, Steen S, Vichiconti J, Posillico D, Cobb M, Medd S, Patel J, Goma S, DiMilia D, Robson M, Duch E, Farinelli M, Wang C, Conti R, Canaperi D, Deligianni L, Kumar A, Kwietniak K, D'Emic C, Ott J, Young A, Guarini K, Ieong M (2005) Enabling SOI-based assembly technology for three-dimensional (3D) integrated circuits (ICs). In: IEEE international electron devices meeting, pp 352–355

107. UKnickerbocker J, SAndry P, Dang B, RHorton R, JInterrante M, SPatel C, JPolastre R, Sakuma K, Sirdeshmukh R, JSprogis E, MSri-Jayantha S, MStephens A, WTopol A, KTsang C, CWebb B, LWright S (2008) Three-dimensional silicon integration. IBM J Res Dev 52(6):553–569

108. Ural A, Li Y, Dai H (2002) Electric-field-aligned growth of single-walled carbon nanotubes on surfaces. Appl Phys Lett 81:34–64

109. Vandevelde B, Okoro C, Gonzalez M, Swinnen B, Beyne E (2008) Thermo-mechanics of 3D-wafer level and 3D stacked IC packaging technologies. In: International conference on thermal, mechanical and multi-physics simulation and experiments in microelectronics and micro-systems, pp 1–7

110. Vardaman J, Garrou P (2010) Global trends in 3D IC packaging. Adv Microelectron 37(3):6–8

111. Wang W, Haruehanroengra S, Shang L, Liu M (2007) Inductance of mixed carbon nanotube bundles. Micro Nano Lett 2(2):35–39

112. Wei B, Vajtai R, Ajayan P (2001) Reliability and current carrying capacity of carbon nanotubes. Appl Phys Lett 79(8):1172–1174

113. Wong E, Lim SK (2006) 3D Floorplanning with thermal vias. In: Design, automation and test in Europe, pp 878–883

114. Wu JH (2006) Through-substrate interconnects for 3-D integration and RF systems. PhD dissertation, Department of EECS, Massachusetts Institute of Technology

115. Wunderle B, Mrossko R, Wittler O, Kaulfersch E, Ramm P, Michel B, Reichl H (2007) Thermo-mechanical reliability of 3-D-integrated microstructures in stacked silicon. In: Materials research society symposium, vol 67, pp 970–974

116. Xie B, Shi XQ, Chung CH, Lee SWR (2010) Novel sequential electro-chemical and thermo-mechanical simulation methodology for annular through-silicon-via (TSV) design. In: 60th electronic components and technology conference, pp 1166–1172

117. Xie Y, Cong J, Sapatnekar S (2009) Three dimensional integrated circuit design: EDA, design and microarchitectures. Springer, New York

118. Y SSMMAKAKDSHITMMOMA (2006) Novel approach to fabricating carbon nanotube via interconnects using size-controlled catalyst nanoparticles. In: 2006 international interconnect technology conference, pp 230–232
119. Yu H, Ho J, He L (2006) Simultaneous power and thermal integrity driven via stapling in 3D ICs. In: IEEE/ACM international conference on computer-aided design, pp 802–808
120. Yu H, Ho J, He L (2009) Allocating power ground vias in 3D ICs for simultaneous power and thermal integrity. ACM Trans Des Autom Electron Syst 14(3):1–31
121. Zhan Y, Zhang T, Sapatnekar SS (2007) Module assignment for pin-limited designs under the stacked-vdd paradigm. In: IEEE/ACM international conference on computer-aided design, pp 656–659
122. Zheng LX, O'Connell MJ, Doorn SK, Liao XZ, Zhao YH, Akhadov EA, Hoffbauer MA, Roop BJ, Jia QX, Dye RC, et al (2004) Ultralong single-wall carbon nanotubes. Nat Mater 3(10):673,676
123. Zhou P, Sridharan K, Sapatnekar S (2009) Congestion-aware power grid optimization for 3D circuits using MIM and CMOS decoupling capacitors. In: Asia and South Pacific design automation conference, pp 179–184
124. Zhu L, Xu J, Xiu Y, Sun Y, Hess DW, Wong CP (2006) Growth and electrical characterization of high-aspect-ratio carbon nanotube arrays. Carbon 44(2):253–258

Biography

Nauman H. Khan received the B.Sc. and M.Sc. degrees in Electrical Engineering from the University of Engineering and Technology (UET), Lahore, Pakistan, in 2002 and 2006, respectively. He received his Ph.D. in Computer Science from Tufts University in 2011 and joined Intel Corporation as a CAD Engineer. Previously, Nauman worked as a lecturer at UET and as a Software Engineer at Techlogix Inc. His research interests include CAD for VLSI systems with particular emphasis on 3D integrated circuits. Nauman has 7 publications in refereed journals and conferences, a book chapter, and a pending patent. He received 3rd place in the 2011 ACM SIGDA Student Research Competition at the Design Automation Conference. He has served as a reviewer for IEEE TVLSI and IEEE ICECS 2010.

Soha Hassoun is with the Department of Computer Science at Tufts University. She earned a Ph.D. from the University of Washington, Seattle, in 1997, and a Master's degree from MIT in 1988. In between degrees, she worked as a circuit designer in the microprocessor design group at Digital Equipment Corporation. She had sabbaticals with IBM's Austin Research Labs in 2002, and with Carbon Design Systems in 2007. Soha's EDA research focus is on understanding how new technologies impact design. Earlier interests include pipelining latency-constrained circuits (PhD thesis) and logic synthesis. Her research interests extend to Bio Design Automation where she focuses on pathway and modularity analysis, synthesis, and predictive modeling. Soha is an NSF CAREER award recipient. She has served on the technical and executive committees for several conferences and workshops including DAC, ICCAD, IWLS, TAU, IWBDA (co-founder), serving key technical roles such as ICCAD's technical program chair (2005) and DAC's technical program co-chair (2011 & 2012). Soha has served as an associate editor for the IEEE Transaction on Computer-Aided Design and the IEEE Design and Test Magazine, and on the Defense Science Study Group. She has served on IEEE's Council on

N. Khan and S. Hassoun, *Designing TSVs for 3D Integrated Circuits*, SpringerBriefs in Electrical and Computer Engineering, DOI 10.1007/978-1-4614-5508-0, © The Authors 2013

Design Automation and as a board member for ACM's Special Interest Group on Design Automation. Soha received several awards from ACM/SIGDA for her service including: the 2000 and 2007 Distinguished Service awards for founding the Ph.D. Forum at DAC, and the 2002 Technical Leadership award. She is a member of ACM and a senior member of IEEE.